FXII
209,50 no

The Diversity of Animal Reproduction

Authors

Richard E. Blackwelder
Benjamin A. Shepherd

Professors
Department of Zoology
Southern Illinois University
Carbondale, Illinois

CRC Press, Inc.
Boca Raton, Florida

Library of Congress Cataloging in Publication Data

Blackwelder, Richard Eliot, 1909—
  The diversity of animal reproduction.

  Bibliography: p.
  Includes index.
  1. Reproduction.   I. Shepherd, Benjamin A.
II. Title.
QP251.B64   591.1'6   81-234
ISBN 0-8493-6355-1   AACR2

This book represents information obtained from authentic and highly regarded sources. Reprinted material is quoted with permission, and sources are indicated. A wide variety of references are listed. Every reasonable effort has been made to give reliable data and information, but the author and the publisher cannot assume responsibility for the validity of all materials or for the consequences of their use.

All rights reserved. This, or any parts thereof, may not be reproduced in any form without written consent from the publisher.

Direct all inquiries to CRC Press, Inc., 2000 N.W. 24th Street, Boca Raton, Florida 33431.

© 1981 by CRC Press, Inc.

International Standard Book Number 0-8493-6355-1

Library of Congress Card Number 81-234
Printed in the United States

PREFACE

There is no possibility of surveying all the groups of animals, for any feature, by direct observation of the organisms. No one person could be adequately qualified for this purpose in more than a small fraction of the 80 or more groups involved. Such a survey cannot even be made from the primary literature, which is even more numerous and scattered than the specimens. It must be done mostly from secondary sources, the treatises and summaries previously published.

If this is not the best conceivable method, it at least seems to be the only practical one. At every step, however, the statements found in the literature must be checked; the implications are often not mentioned; the facts may be oversimplified or misinterpreted. After making adjustments for all of these, the diversity that remains in animal reproduction is far greater than admitted by any book we have seen. It has far-reaching implications for zoology and especially genetics. It will make classification more difficult (at least for a time), and it will make phylogenetic speculations less certain, at many levels.

All this because exceptions to the usually used generalizations are legion. When we see a statement that in a certain class the feature under discussion is usually present, or that it generally develops by a certain sequence, or that it is cyclic in most species, the words *usually, generally,* or *in most* tend to slip by us without registering in our minds. Such a statement appears to be an attempt to hide the diversity. It can give only a warped view of the capabilities of the group in question; it can lead to false conclusions about the species in hand; and it can lead to false expectations about the genetic make-up of every population.

In gathering our information from many books and articles, it is not supposed that the cited writers were the discoverers of the information they report; they are merely our immediate sources. Much of what is here ascribed to a certain source author was not even directly stated by him but can be inferred from statements or figures given by him. For example, most writers refer to polyembryony as a developmental process, but it is surely also reproductive and is so listed here. Again, many writers accept self-fertilization and parthenogenesis as part of sexual reproduction, but these processes do not give the genetic result expected from "real" bisexual (karyogamic) reproduction.

It is our purpose to bring together what is known of the reproductive occurrences in all groups of living animals, to show the total diversity, the almost random distribution of the processes, and the implications for the rest of zoology. All processes are herein cited under the terms as we define them.

Completeness is unattainable. First, we have no doubt missed information already published; second, for many animals there is no certain knowledge of the reproductive methods. It is thus likely that every specialist will be able to cite further examples, more diversity, and corrections to our accumulation. We consider it inevitable, however, from our experience so far, that the final result of these emendations will be to increase further the diversity that can be listed.

This book is primarily a monograph of the reproductive diversity among animals, including protozoans. This diversity is listed for each group in Chapter 6; it is cross-listed by process in Chapter 7. In order to use terms uniformly, they had to be defined (Chapter 4 and Glossary), but in many cases the decision on a definition required discussion (Chapters 2 and 3). The additional complexities that may be found in the occurrence of more than one type of reproduction in a given species are suggested in Chapter 5.

Richard E. Blackwelder
Benjamin A. Shepherd

THE AUTHORS

Richard E. Blackwelder was born in 1909 and received his Ph.D. from Stanford University in 1934. For 20 years he was associated with the Smithsonian Institution, as holder of the W. R. Bacon Travelling Scholarship and as Associate Curator of Insects. In 1956 he went to Southern Illinois University, broadening his interests to include all animals. He taught Zoology, Invertebrates, and Taxonomy, always emphasizing the extent of the diversity of animals. He retired from his Professorship in 1977. He was an officer of the Society of Systematic Zoology for 16 years from its founding and has published over 150 titles, including a dozen books.

Benjamin A. Shepherd is Professor of Zoology and Associate Vice President for Academic Affairs (Services) at Southern Illinois University at Carbondale, Carbondale, Illinois.

Dr. Shepherd received the B.S. degree from Tougaloo College, the M.S. degree from Atlanta University and the Ph.D. from Kansas State University. He has trained several graduate students in the Department of Zoology at Southern Illinois University and published research findings in a variety of journals. His research interests have been primarily in the area of reproductive physiology of the male guinea pig. He is a member of a number of learned societies and was selected as an American Council on Education Fellow in 1978.

ACKNOWLEDGMENTS

This study was begun about 12 years ago by the senior author and James K. V. Adams, then a student at Southern Illinois University. Mr. Adams has not been available to work further on the manuscript, which has recently been enlarged and rewritten as part of an even larger study of diversity of animal processes and structures. Mr. Adams' contribution is gratefully acknowledged, but he has escaped responsibility for any shortcomings.

Our other major acknowledgment is to the authors who assembled the facts that are here summarized and tabulated. These are listed as References. Among them, especially useful were the works of Hyman, Vorontsova and Liosner, Kaestner, Kume and Dan, and Barnes.

TABLE OF CONTENTS

Chapter 1
Introduction 1
- I. Difficulties 1
- II. "Process" Defined 3
- III. Implications 3
- IV. Axioms Basic to this Study 4
- V. Diversity Factors 5
- VI. General Conclusions 5

Chapter 2
Some Basic Concepts 7
- I. Introduction 7
- II. Individuality 7
- III. Sexuality and Sex 9
- IV. Reproduction 11
- V. Reproductive Bodies 14

Chapter 3
Sexual vs. Asexual 17
- I. Introduction 17
- II. The Processes of Reproduction 17
  - A. Sexual vs. Asexual 17
  - B. Bisexual vs. Unisexual 18
  - C. Gametic vs. Nongametic 19
  - D. Amphimictic vs. Apomictic 19
  - E. Meiotic vs. Ameiotic 20
  - F. Meiosis and Genomes 20
- III. Bisexual Reproduction 21
- IV. Basic Bisexual Reproduction 22

Chapter 4
The Many Terms and Processes 25
- I. The Reproductive Processes 25
  - A. Fertilization and Activation 25
  - B. Parthenogenesis 28
  - C. Asexual Reproduction 30
- II. Parareproductive Processes 37

Chapter 5
The Levels of Diversity 39
- I. Introduction 39
- II. Sequences of Cycles 39
- III. Reproductive Processes in a Species 46
  - A. Single Segment Sequence 46
  - B. Multiple Segment Sequence 48
- IV. Diversity Within a Class 49

Chapter 6
Reproduction in Animals by Class 57
- I. Introduction 57

II.	Phylum Protozoa	57
	A. Class Sarcodina	57
	B. Class Flagellata	58
	C. Group Opalinda	59
	D. Class Ciliata	59
	E. Class Suctoria	60
	F. Class Sporozoa	60
III.	Phylum Porifera	61
	A. Class Calcarea	61
	B. Class Hexactinellida	62
	C. Class Demospongia	62
	D. Class Sclerospongea	62
IV.	Phylum Mesozoa	63
	A. Class Dicyemida	63
	B. Class Orthonectida	63
V.	Phylum Monoblastozoa	64
VI.	Phylum Placozoa	64
VII.	Phylum Coelenterata	64
	A. Class Hydrozoa	65
	B. Class Scyphozoa	66
	C. Class Anthozoa	67
VIII.	Phylum Ctenophora	68
	A. Class Tentaculata	68
	B. Class Nuda	68
IX.	Phylum Platyhelminthes	68
	A. Class Gnathostomuloidea	69
	B. Class Nemertodermatida	69
	C. Class Xenoturbellida	69
	D. Class Turbellaria	69
	E. Class Temnocephaloidea	70
	F. Class Trematoda	70
	G. Class Cestoda	71
	H. Class Cestodaria	72
X.	Phylum Rhynchocoela	72
XI.	Phylum Acanthocephala	73
XII.	Phylum Rotifera	73
	A. Class Seisonidea	73
	B. Class Bdelloidea	73
	C. Class Monogononta	73
XIII.	Phylum Gastrotricha	74
	A. Class Macrodasyoidea	74
	B. Class Chaetonotoidea	74
XIV.	Phylum Kinorhyncha	74
XV.	Phylum Priapuloidea	75
XVI.	Phylum Nematoda	75
XVII.	Phylum Gordioidea (Nematomorpha)	75
	A. Class Gordioidea	76
	B. Class Nectonematoidea	76
XVIII.	Phylum Calyssozoa (Entoprocta, Endoprocta, Kamptozoa)	76
XIX.	Phylum Bryozoa (Ectoprocta, Polyzoa)	76
	A. Class Phylactolaemata	76
	B. Class Gymnolaemata	77

XX.	Phylum Phoronida	77
XXI.	Phylum Brachiopoda	78
	A. Class Inarticulata	78
	B. Class Articulata	78
XXII.	Phylum Mollusca	78
	A. Class Monoplacophora	78
	B. Class Amphineura	79
	C. Class Solenogastres	79
	D. Class Gastropoda	79
	E. Class Bivalvia	79
	F. Class Scaphophoda	80
	G. Class Cephalopoda	80
XXIII.	Phylum Sipunculoidea	80
XXIV.	Phylum Echiuroidea	80
	A. Class Echiurida	81
	B. Class Saccosomatida	81
XXV.	Phylum Myzostomida	81
XXVI.	Phylum Annelida	81
	A. Class Polychaeta	81
	B. Class Oligochaeta	84
	C. Class Hirudinea	84
	D. Class Archiannelida	84
XXVII.	Phylum Dinophiloidea	85
XXVIII.	Phylum Tardigrada	85
	A. Class Heterotardigrada	85
	B. Class Eutardigrada	85
XXIX.	Phylum Pentastomida	86
XXX.	Phylum Onychophora	86
XXXI.	Phylum Arthropoda	86
	A. Class Merostomata	86
	B. Class Pycnogonida	87
	C. Class Arachnida	87
	D. Class Crustacea	87
	E. Class Pauropoda	88
	F. Class Symphyla	88
	G. Class Diplopoda	88
	H. Class Chilopoda	88
	I. Class Insecta	88
XXXII.	Phylum Chaetognatha	89
XXXIII.	Phylum Pogonophora	89
XXXIV.	Phylum Echinodermata	90
	A. Class Crinoidea	90
	B. Class Somasteroidea	90
	C. Class Asteroidea	90
	D. Class Ophiuroidea	91
	E. Class Echinoidea	91
	F. Class Holothurioidea	92
XXXV.	Phylum Pterobranchia	92
XXXVI.	Phylum Enteropneusta	93
XXXVII.	Phylum Planctosphaeroidea	93
XXXVIII.	Phylum Tunicata	93
	A. Class Larvacea	93

	B.	Class Ascidiacea	93
	C.	Class Thaliacea	94
XXXIX.		Phylum Cephalochordata	94
XL.		Phylum Vertebrata	95
	A.	Class Agnatha	95
	B.	Class Chondrichthyes	95
	C.	Class Osteichthyes	95
	D.	Class Amphibia	95
	E.	Class Reptilia	96
	F.	Class Aues	96
	G.	Class Mammalia	96

Chapter 7
Distribution of Processes 97

Appendix 1
Glossary 105

Appendix 2
Classification and Annotation 119

References 131

Index 133

Chapter 1

INTRODUCTION

It has been said many times that among all the processes and activities of organisms, reproduction is the most important. The reasons for this statement are obvious, but it is also obvious that all essential processes — those necessary to life — are equally important, and there are many other processes that meet this requirement. Nonetheless, reproduction does occupy a unique place in biology, because it is central to heredity and thus to evolution. At the present time there is an additional reason for studying reproduction at this level of individual replacement, and this is the widespread misunderstanding of the extent of the diversity among these processes.

The word reproduction is not one to send many readers to a dictionary. It readily conveys the idea of production of additional individuals of that species, either singly or in large numbers, depending on the previous experience of the reader among the various groups of animals. It may sometimes raise visions of diversity in the accompanying behavior, but bisexual reproduction is usually assumed, even though the existence of asexual reproduction among "lower animals" is recognized.

I. DIFFICULTIES

Unusual processes have heretofore been discovered in many groups of animals and many terms have been coined to describe them. An attempt to bring together these diverse processes and to give an overview of all reproduction in all animal groups soon runs into difficulties of several sorts. Six of these difficulties are summarized here.

First, the fact that most zoologists are more familiar with vertebrates than with invertebrates makes it likely that conditions in the vertebrates will seem to be the norm or even considered to be universal. It is seldom realized how erroneous these assumptions are. Although some form of sexuality occurs in most groups of animals, there are species in several in which sex is unknown. In approximately half of all animal species there occurs some form of nonbisexual reproduction that interrupts the lives of the bisexually produced individuals. Probably a very large majority of all individual animals were themselves produced nonsexually, even if bisexual reproduction occurs elsewhere in the species.

Second, in vertebrates most of the obvious reproduction occurs in the adult state, which is defined on that basis. But in animals in general there is probably more reproduction in embryonic and larval stages than in the "adult." These stages are almost exclusively asexual in their reproduction.

When there is reproduction in early developmental stages, it is likely to be passed over as a developmental process. Polyembryony and budding and fragmentation of embryos and larvae do produce new individuals. It is impossible to deny that they are reproductive, even if they are usually studied by developmentalists rather than by those specializing in reproduction.

Third, most zoologists learn early the developmental sequence of higher vertebrates, involving a fertilized egg, multiplication and diversification of cells, and development through embryonic or larval forms into adulthood. This sequence is generally believed to be followed by all individuals of each vertebrate species and most other animals. However, among all animals, many millions of individuals of thousands of species are produced entirely without this zygote-to-adult developmental sequence, or at least by shortcutting parts of it. A wide variety of species in many groups are incapable of the usual sexual processes and reproduce solely by what are sometimes considered "exceptional" means. These exceptional means are less rare and more widespread than is generally realized.

In addition, even more species involve a cycle of greater complexity in which there are several episodes of reproduction, often by processes that produce numerous offspring simultaneously. Such is the multiple fission in Protozoa, the multiple "budding" in tapeworm larvae, and the embryonic fission that produces thousands of parasitic Hymenoptera from one fertilized egg. These cycles actually involve a sequence of individuals, with most of them not reproducing sexually.

Fourth, in many species the manner of production of one individual may be quite different from that of other individuals. One may be produced by cross fertilization and a second by parthenogenesis (as in aphids), or one by cross fertilization and a second by one of the asexual processes (as in polyembryony), or one by parthenogenesis and a second by one of the asexual processes, or one by one of the asexual processes and a second by another. (These are tabulated in detail in Table 3 in Chapter 3.)

In the reproductive cycle of certain animal species, both sexual and asexual processes occur. The latter tend to be overlooked when both sexes are present, with the reproduction assumed to be outbreeding and thus to give expected Mendelian ratios. In this manner a normal bisexual reproduction may be followed in the larval stage by an asexual process, such as strobilation (as in Scyphozoa), to produce many individuals that are genetically identical. These resulting individuals form a clone, because they were not produced by the karyogamy which would be necessary to yield the expected character frequencies.

Nonbisexual reproduction may occur anywhere in the cycle. This means not only as adult (parthenogenesis, budding, fragmentation) or as larva (budding or parthenogenesis), but also as embryo (budding, successive fragmentation) or even as zygote (polyembryony).

It seems to be frequently assumed that a particular cycle is bisexual — amphimictic — without making certain that apomictic processes do not replace the amphimixis or intervene in the cycle. In the literature, bisexual reproduction in the strict sense (herein called Basic Bisexual) is often hard to identify because (1) there is usually no direct reference to karyogamy, (2) parthenogenesis may occur unrecognized because it may occur even in dioecious animals in which individuals are known to come together in sexual behavior, and (3) some apomictic processes are regarded as developmental rather than reproductive by some writers and so are not mentioned.

To be certain that it is reasonable to assume that a given species will produce expected genetic ratios in its members, it is necessary to be certain that there are no apomictic processes, no unisexual processes, no self-fertilization, and no clonal fertilization in the cycles of any of the reproducing members of that species. This leads us to a more restricted concept, described below.

Fifth, the number of terms employed for the various processes of reproduction is very large; over 500 are listed in our Glossary. There are many subprocesses (e.g., arrhenotoky, thelytoky, deuterotoky) in parthenogenesis and many synonyms (e.g., outbreeding, exogamy, allogamy). Even without these, there are at least 35 separate forms of reproduction (as listed at the top of page 3), but these are combined into at least 250 different sequences in the various animal groups. All of these are discussed in later sections.

Sixth, when the term sexual reproduction is used to cover all reproduction by animals that produce gametes, there are a variety of unnoticed processes whose implications are omitted. These include several kinds of parthenogenesis, self-fertilization, pseudogamy or plasmogony, and autogamy. These are each said to be sporadic in occurrence, but in some

Gamogony	Agamogenesis	Plasmotomy (continued)
Anisogamy	Sporogony	Successional
Isogamy	Gemmulation	polyembryony
Hologamy	Sorites	Budding
Staurogamy	Statoblasts	Strobilation
Exogamy	Podocysts	Architomy
Endogamy	Hibernacula	Pedal laceration
Mychogamy	Fission & division	Frustulation
Parthenogenesis	Tomiparity	Autotomy
Pseudogamy	Schizogony	Scissiparity
Plasmogony	Merogony II	Fissiparity
Gonomery	Plasmotomy	Paratomy
Merogony I	Polyembryony	Stolonization
		Epitoky
		Eudoxy

species in most groups of animals at least one of these reproductive processes is obligate. The species with no known true bisexual reproduction are more numerous than usually recognized. They occur in nearly all phyla of animals but are frequently overlooked.

When reproductive processes are distinguished as sexual and asexual, there is usually a substantial "no man's land" between where would be found, unmentioned, self-fertilization, and very often parthenogenesis. For this reason it is ineffective to classify reproductive processes as "sexual" and "asexual." Similarly, the terms "amphimixis" and "apomixis" do not clarify this situation, since amphimixis includes only sexual processes whereas apomixis includes both sexual and asexual ones. This duality of the latter makes it necessary to subdivide apomixis into sexual apomixis and asexual apomixis. (These difficulties are discussed further in Chapter 3.)

II. "PROCESS" DEFINED

It is impossible to discuss reproduction or any other activity of living things without using certain words that have no clear single meaning. For example, the word "process" is extremely useful to denote the various methods by which animals accomplish each of their activities. It is at once seen, however, that the processes are at many levels, often with one process consisting of several others at a lower level. Thus, reproduction can be called the process which produces new individuals, but we know that reproduction can involve the process of fertilization which itself can involve the process of gametogenesis which itself involves the process of meiosis which itself involves the process of synapsis and so on. There seems to be no way to avoid a slight lack of specificity in this word, except that it can sometimes be coupled with "system," "sequence," or "method" to remind us of the breadth of meaning it can carry. We will use it in many situations and merely try to avoid ambiguity without excessive circumlocution.

III. IMPLICATIONS

The unity of all life has become more obvious in the past few decades as the RNA-DNA systems, genes, amino acids, proteins, and other features have come to be recognized as universal among organisms. This unity has been greatly emphasized in recent books. It is not usually noted that these features are themselves diverse among the organisms.

From the viewpoint of some zoologists, the similarities — unity — seem to be dominant. When the diversity of detail and process are examined, however, the general unity may seem to be primarily the result of our habit of generalizing.

This happened with the Theory of Recapitulation, which at one time seemed to promise to unify all animals in an unfolding system that could be positively demonstrated. The diversity of development has long since shown this early promise to be only a dream, even though some aspects of recapitulation theory may still be useful.

It also happened with some generalizations of development, which once promised clear explanations of how an egg or embryo differentiates into individuals of that species. The diversity in early development, such as the first few cleavages and even gastrulation, has proven to be so great that the generalizations have had to be replaced by separate descriptions in many cases.

More recently it has been held that the laws of Mendel apply to all organisms. As investigations continued, numerous situations and occurrences were found that forced additions to Mendel's laws and added supplementary laws. It is often forgotten that scarcely a thousandth part of the species of animals have been studied at all genetically. Some of the uniformities that have been taken for granted — meiosis, gametes, zygotes, recombination — are now known to be lacking in some species. There is likelihood that even more diversity in genetic mechanisms will be discovered, as has recently occurred in sex determination.

Implications of bisexual reproduction in various aspects of biology (genetics, evolution, behavior) are so great and of such importance that some biologists feel that any other form of reproduction is secondary, less fundamental, or less worthy of study. In these fields, however, it is not sexuality itself that is important but outbreeding, the mixture of genetic materials from two individuals. Few biologists have taken note of Sonneborn's estimate that probably half of all animal individuals were the direct result of nonoutbreeding reproductive processes. This large number cannot reasonably be labeled as exceptional. One of the purposes of the present catalogue of diversity is to show how frequently the sexual reproduction is obscured or replaced by other processes.

IV. AXIOMS BASIC TO THIS STUDY

The diversity of reproduction tabulated in this study has forced us into a series of decisions and definitions that have to be understood if the reader is to follow the conclusions. The following are axioms basic to the interpretation that we adopt:

1. New individuals may arise by (a) reproduction, in which additional individuals are produced, (b) parareproduction, in which the nuclei of one or two individuals are genetically changed without an increase in the number of individuals (as in conjugation), or (c) fusion of two or more to form a single new one.
2. Reproduction occurs only in an individual or from parts of two individuals (or in individual colonies).
3. Every reproduction produces a new individual. This can be by successive acts of reproduction by one individual or one pair, in each case of which a new individual is produced or it can be by a succession of different processes, each of which produces a new individual whether or not the original parent survives. There is thus only one reproductive process in the background of each individual. These individuals include the (a) one that is produced by each bisexual process, (b) the one that is produced by each unisexual process, (c) the two that are produced by binary fission or some form of fragmentation, or (d) the many produced by multiple fission or fragmentation.
4. An individual exists from the moment of fertilization (or activation) of an ovum or

from the start of independent activity in a fragment or a bud or the fusion of two or more individuals.
5. In several asexual processes the parent ceases to exist as an individual, no matter how brief the time involved since its formation.
6. An individual may survive past a reproductive episode, but if its genome is changed it achieves a new individuality. Likewise if all its "nuclear material" becomes incorporated into new individuals (as in multiple fission), it ceases to exist (i.e., its individuality is lost).
7. There are species in which there occurs a succession of reproductive processes, often alternating sexual and asexual, but no individual passes through two reproductive phases except in the sense of repeated episodes of the same process.
8. There are very many species in which there exists more than one pathway of reproduction. Some individuals use one pathway (series of processes) and others another.
9. Reproduction is not a necessity for any individual and may not even be possible, but some individuals in every species must reproduce if the species is not to become extinct.

V. DIVERSITY FACTORS

Reproduction is so extremely diverse among animals because of these factors:

1. There are more than a dozen discrete methods in more than two score of sequences that may be employed.
2. Most of these can occur at any stage in the life of the animal or in several stages of it.
3. Within a given species there may be as many as four different reproductive pathways employed by different individuals.
4. Some individuals exist from fertilization through adulthood to death, but others may cease to exist at much earlier stages because they give rise to two or more new or different individuals by the disappearance or transformation of themselves.
5. Sexual processes are not restricted to what is usually called sexual reproduction.
6. In addition to individuals that can be recognized, there are composite aggregates consisting of parts of many individuals. There are pairs of individuals that exchange genotypes. There are fused individuals that never separate. There are living masses formed by fusion of many individuals. There are colonies in which most individuals have disappeared in the production of the highly individualistic colony. Any of these can reproduce themselves. These make it impossible to apply *any* definition of individual rigidly.

VI. GENERAL CONCLUSIONS

The compiling of the data presented hereinafter on the diversity and distribution of reproductive processes has lead to several conclusions. First, the extremely scattered information on many reproductive processes has led to general failure to recognize how inadequate the discussions of them are and how shaky the genetic assumptions that are (often unconsciously) drawn. Second, the available terms are inadequate to classify the processes, which are diverse in so many ways that no one classification serves many interests. Third, the factors and modes of sexuality, individuality, and reproduction need to be defined for all animals at once, not merely to the satisfaction of those interested in any one. Fourth, the most basic consideration in classifying or understanding reproductive processes is meiosis, its occurrence, and its extent. Fifth, reproduction-related processes include not merely multiplication of individuals but also some prereproductive processes and those that produce new individuals by genetic change only (parareproductive). Sixth, the distribution of processes in the different phyla may be related to their manner of living, but it gives no support to any postulated phylogeny, giving much the same appearance as random distribution.

Chapter 2

SOME BASIC CONCEPTS

I. INTRODUCTION

Four ordinary words that are generally used in special senses in later chapters are defined here. They are Individuality, Sexuality, Sex, and Reproduction. The definitions do not conform to common usage because the common usage has been found to be inadequate when the full range of diversity in animals is considered. It therefore seems to be necessary to give here a discussion of why these particular definitions have been chosen. An unusual way of viewing all reproduction as arising from body fragments of some sort is also given, under the heading "Reproductive Bodies."

II. INDIVIDUALITY

In order to discuss the forms of reproduction in depth, it is necessary to refer to individuals. Individuality is familiar enough among vertebrates, but many difficulties arise when we open our vista to include all animals. The diversity of animal life cycles may be reflected in complexities in the reproductive sequences. Animals often occur in several forms in a species, differing in their reproduction. They may occur as aggregates with no clear individuality, as in some sponges; in colonies with definite form to the colony but the "individuals" only partly distinct, as the siphonophores; or as composite structures in which many "individuals" share some organs, as in ascidians. There is no possibility of clear-cut distinction between individual and colony in these circumstances.

Exclusive definition of individual is impossible because of the diversity that exists. Among the difficulties encountered as a result of the diversities are the following:

1. Every individual passes through a series of stages in its life history. If a grown man is an individual, was he also one as a child? As an unborn foetus? As a zygote?
2. In colonial animals, does the bud become an individual only on detachment? Do colonies of attached zooecia consist of individuals? When a hydra produces a lateral bud, at what point does this become an individual?
3. Does it make a difference how the new tissue arises? The parental tissues and organs may produce the new individual by fission with little further growth or differentiation, or single cells from several tissues may assemble into the new fragment, which must then grow and differentiate (gemmulation). Do these both produce new individuals?
4. Are the following to be considered new individuals when isolated: (1) isolated gametes, because they are alive, (2) the fragments of a starfish resulting from autotomy when both are capable of regenerating into whole starfish, (3) detached proglottids of tapeworms, and (4) a spermatophore, or a hectocotyle?
5. Does loss of form involving drastic reorganization of tissues and cells followed by re-formation produce a new individual or the same one? When a tardigrade dedifferentiates to a formless mass and then grows and redifferentiates into an adult again, is it the same individual or a new one?

Such questions can be asked almost *ad infinitum,* and they might seem to require a complex definition of the word individual. This would be necessary if one requires a completely objective definition that leaves no doubts in any theoretical case, but such finality seems unattainable with the diversity reported here. The problems that we have faced in arriving at a working definition are discussed below.

First, when in development does a new individual come into existence? The moment at which individuality is achieved varies from the instant of separation (in binary fission) to the instant of fertilization (in karyogamic species) and to the assumption of the first inherent function (in a bud). The moment cannot be pinpointed in the case of parthenogenetic development of an ovum because activation cannot be observed in the absence of karyogamy. In the case of polyembryony, the individual produced at fertilization will not be the same as the two or more individuals resulting from the following fragmentation. There are species in which embryos form inside "unborn" embryos and tertiary embryos inside each of these. These are all individuals by our definition because they are organized bodies in the characteristic form for that stage of that species and thus are each the start of a new life cycle. (See *Gyrodactylus* in Chapter 6, Section IX. F. Table 2.)

Thus, a zygote is an individual and so is an embryo, larva, an adult, a mature bud, a regenerating fragment, or an activated ovum. Such a new individual can arise by any of these events:

1. Activation of an ovum by (a) fertilization or (b) by other activation
2. Fragmentation (in the broad sense) — agametes, buds, gemmules, fragments
3. Fusion of blastomeres or larvae or unicellular adults

At the same time individuals can cease to exist through any one of three events:

1. Physical death, natural or accidental
2. Fission in which all the tissues (or cell parts) are passed on to daughter individuals
3. Fusion of two individuals into a new one (hologamy)

Second, to what extent do two individuals have to be separated? They must be separated physically or functionally and perform at least some functions independently.

Third, are all detached fragments new individuals? No. Detached fragments are new individuals only if they are fragments capable of complete regeneration and performance of some of the functions characteristic of that species.

Fourth, is it possible to distinguish developmental processes from reproductive? No. They can be looked upon as either or both, depending upon the viewpoint involved. In the case of polyembryony, for example, if one considers a single offspring, the fragmentation was one of the processes in its development; but when one considers what happens to the zygote, it clearly produces fragments that develop separately. The fragmentation was a reproductive process, and therefore all subsequent individuals originated from the zygote by reproduction.

Fifth, plasmodia, masses of protoplasm with nuclei but no cell separations, remain equivocal as to individuality.

Sixth, many sorts of colonies have individuality, as in *Volvox* and *Physalia*. In these there seem to be no sharp distinction between reproduction by an individual and reproduction by a colony, which prevents us from determining the individuality of the so-called "persons" of which the colony is composed. In view of the considerations above, we consider as individuals all of these living objects:

1. Every zygote—and the single animal which develops from it
2. Every unfertilized ovum which is activated rather than fertilized and is capable of development—and the single animal which develops from it

3. Every agamete capable of development—and the single animal which develops from it
4. Every cell formed by fission (binary or multiple) of a unicellular individual, whether or not separation is complete
5. Every multicellular body formed by fragmentation, fission, autotomy, or polyembryony, if capable of normal development and independent existence
6. Every body formed by proliferation of tissue (budding or strobilation), if capable of normal existence whether separated or remaining attached
7. Every new body formed by incorporation of cells from various tissues, if capable of growth and independent existence (gemmules, statoblasts, etc.)

Living objects which would not be considered to be individuals would include spermatophores, ova which have not been fertilized or activated, spermatozoa under any circumstances, tapeworm proglottids, and autotomized appendages that lack the ability to regenerate. We think it appropriate to sidestep the question of individuality or coloniality of sponges and plasmodia, as these are out of the range of either concept. (A tardigrade reduction body is the regressed body of one individual, when it is rebuilt it is neither a new individual nor a product of new reproduction.)

The diversity of completely separate animals is only part of the diversity of individuality because other animals may exist with only partial separation, as in colonies. Animals exist not only as single cells or as multicellular complexes (metazoans), but also as colonies of single cells and colonies of the multicellular complexes and even as composite structures consisting of "half" individuals sharing other organs (as in some Thaliacea).

An individual is an organized entity with the potential of performing integrated activities necessary to its continued existence. The entity may be either a protozoan or a metazoan. In Protozoa it is a cell (or in a plasmodium a mass of protoplasm under the influence of its nucleus). In Metazoa it is an aggregation of cells, cell products, and syncytia.

III. SEXUALITY

Sex is the phenomenon of members of a species existing in two forms, generally identified as male and female. In such species, it is assumed that one member of each "sex" will contribute a haploid complement of hereditary factors to the next generation, but this may not occur.

Sexuality is then the state of organisms exhibiting sex. Although most animals show some evidence of sex, there are some that do not. These latter are neuters; they reproduce only asexually, if at all.

The dimorphism upon which maleness and femaleness are based may include differences in (1) sex chromosomes, (2) appearance of gametes (cytological), (3) anatomical and/or morphological characteristics of the animal as a whole or of specific structures, (4) behavior, or (5) subtle physiological differences. Sex is thus fundamentally dual, but in many species one of the sexes is absent (always the male). Thus sexuality is shown by the males and females in some species, by the hermaphrodites in others, and by the females alone in still others.

Here we are using "sexual" to describe all reproduction-related processes that are not strictly asexual. In general, this includes gamogony, parthenogenesis (when meiotic), hologamy, conjugation, nuclear reorganization, and associated behavior. Not all of these are reproduction but merely involve the features we call sex. Later paragraphs will further classify these processes and define them.

Sexual individuals are those that can take part in some form of sexual reproduction, and these can only be males, females, or hermaphrodites. There are also individuals that do not reproduce sexually but have some of the features of such sexual individuals, these include some of the neuters as well as individuals in which the two sexes are mixed (gynandromorphs and intersexes).

A species is said to be **dioecious** if it contains separate male and female individuals. The condition is **dioecism**. It is said to be **monoecious** if there are only individuals with both sexes, in which case the condition is **monoecism**. An individual is said to be gonochoristic if it has only the organs of one sex, and this is **gonochorism**. An individual is **hermaphroditic** if it contains both types of gonads, either at the same time or at different times, and this condition is **hermaphroditism**. There are a few species where three types of individuals occur, males, females, and hermaphrodites, but there seems to be no term to indicate this.

Hermaphroditic individuals therefore have the ability to produce both spermatozoa and ova. These gamete-producing abilities may occur in one of the following sequences.

A. Simultaneously bisexual — This produces spermatozoa and ova simultaneously (1) in separate gonads with separate ducts and no self-fertilization, (2) from a single gonad and therefore with self-fertilization possible or (3) from two gonads that have a common duct and therefore with self-fertilization possible (See Temnocephaloidea, Section IX. F. in Chapter 6).

B. Single change of sex — This produces (1) spermatozoa first and then ova, from an ovotestis, (2) spermatozoa first and then ova from separate gonads, or (3) ova first and then spermatoza.

C. Alternating sex changes — This produces gametes of each sex alternately.

Hermaphroditism occurs in 51 metazoan classes, as listed in Chapter 6. The situations related to hermaphroditism are

1. Sexes permanently in separate individuals (gonochorism)
2. Occasional pathological sex reversal in an individual
3. Normal reversal of structural sex in the life cycle: male first (protandry) or female first (protogyny)
4. Permanent union of sexes in one individual (a) gametes maturing at same time (synchronogamy) or (b) at different times (dichogamy), with the male first (protandry) or the female first (protogyny); (c) gonad transplant; (d) gynandromorphism (pathological condition in which the individuals are half male and half female); and (e) intersexuality (pathological condition in which the sexual features are mixed)

Special terms relating to hermaphroditism are

Syngony — This is sometimes used in reference to Nematoda as synonymous with hermaphroditism.

Synchronogamy — This is the condition in hermaphrodites of having the male and female sex organs functional at the same time. This would exist among hermaphrodites wherever protandry or protogyny do not occur.

Dichogamy — This is the condition in hermaphrodites of having the gonads functional at different times. (Compare synchronogamy and sex reversal.) It occurs as protandry or protogyny.

Protandry — This is the condition in which an hermaphrodite produces male gametes first and later female.

Protogyny — This is the condition in which an hermaphrodite produces female gametes first and later male.

Sex reversal — This expression would literally include all sequential hermaphroditism, but it is usually reserved for the occasional pathological (or at least unusual) change of sex.

Gonad transplant — This occurs in one species of Crustacea, where at any early age the male injects into the female body cavity (by hypodermic injection) a few dedifferentiated cells. These then develop into a gamete-producing testis which makes the female ostensibly hermaphroditic. The zygotes are apparently produced by "self-fertilization," but genetically they are the result of normal cross fertilization.

Sex arrangement in a species — Sex can appear in a variety of combinations, including these:

1. Males and females with testes and ovaries, respectively
2. Hermaphrodites with both testis and ovary
3. Males, females, and hermaphrodites
4. Males and hermaphrodites
5. Females only (parthenogenesis)
6. Males only (This cannot occur except if the individuals are reproductively neuter, reproducing asexually. There are also a few species known only from males, but females will presumably be found.)

It should be noted that in (1) and (3) above there may also be neuter (nonsexual) individuals. Furthermore, there are species in which only neuters are known, all reproduction being asexual. In (1) and perhaps others there may also occur gynandromorphs or intersexes, and in Protozoa there are physiological forms called mating types (two or more per species) which will conjugate with members of different types but not with members of their own type. Other terms applied to sexual reproduction are **gamogenesis, zoogamy, zygogenesis,** and **amphigony.**

IV. REPRODUCTION

The word **reproduction** is often used as if it were a single process acting in a single organ system. In reality, it is always a complex of processes, often extending through the life of the individual. In general, it covers all the processes which lead to production of new individuals; this is substantially equivalent to the life of the individual because there is little in that life that does not contribute to reproduction. Specifically, **reproduction is the origination of new organisms from preexisting ones,** which in fission is direct and immediate but in bisexual reproduction involves operation of gonads and many associated organs, resulting in two sets of gametogenesis, delivery of the spermatozoa to the ova, and fusion of the two.

Reproduction has been defined also as the act of a living system giving rise to new systems identical with itself, but it must not be forgotten that frequently reproduction will produce offspring very different from the parent (and not capable of development into one), other reproductive processes intervening before an individual similar to the original parent appears. For example, a hydroid polyp such as that of *Bougainvillea* may bud another polyp like itself or it may bud a gonophore which must develop into a medusa which must produce gametes to be fertilized before the sequence can produce a separate polyp similar to the original one. Likewise, in a colonial animal such as *Obelia,* the hydranth polyp can bud other hydranths like itself or it may bud a gonangium polyp which must bud medusae which must produce gametes to be fertilized before the sequence can produce a separate hydranth again.

Even in somewhat higher animals this can be illustrated. In Insecta such as aphids

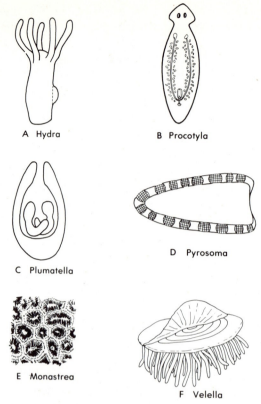

FIGURE 1. Diagrams of multiple or compound individuals. (A) solitary polyp, (B) hermaphrodite, (C) dual embryo, (D) compound structure, (E) encrusting colony, (F) individualistic colony.

where parthenogenesis is obligate, the sexually produced females give rise by parthenogenesis to individuals distinctly different from themselves. In some Polychaeta, epitoky may give rise asexually to a quite different "epitoke" which then reproduces sexually to form another like the original parent.

There are several ideas or conditions that make it difficult to discuss the diversity of reproduction. One is that reproduction is usually defined as production of new "individuals," although there is no satisfactory way to define that word. Another is the existence of colonies with great individuality, of sexually dual individuals such as hermaphrodites (Figure 1B), of compound individuals (Figure 1D), and of three types of ova visibly indistinguishable: (1) haploid which must be fertilized; (2) haploid which develops without fertilization; and (3) diploid which also develops directly.

Sexual reproduction is often thought of as universal, and some sexual process does occur in some form in most groups of animals. Yet there has been no simple way to say what is sexual and what is not. Sexuality is a condition that may exist independently of any actual reproduction, and much so-called sexual reproduction does not clearly involve sexuality at all. The examples of the latter include all instances of ameiotic parthenogenesis, usually tacitly included in sexual reproduction, in which the parent may have features of a female, but where there is in the reproduction neither amphimixis nor meiosis.

Any individual may fail to reproduce or may even be incapable of reproduction. Thus, at the individual level, reproduction is not a necessity; there are thousands of species in

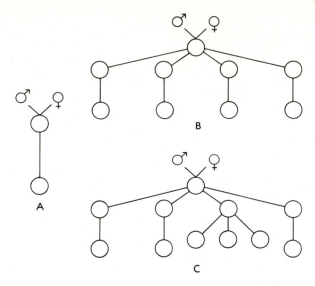

FIGURE 2. (A) The numerical outcome in Basic Bisexual Reproduction and (B) and (C) in a species showing polyembryony. In (A) a single cycle is shown. In (B) and (C) each diagram spans the life of two (or three) successive generations, separated at the moment of polyembryony (which occurs as successive fragmentation).

which most individuals never reproduce in any way. It is at the level of species that reproduction becomes essential. If biologists are right in believing that a given species can evolve only once, then it follows that no species can continue to exist without replacement of dying individuals.

When we speak of reproduction, we do not usually mean a single process but a series of processes, because there is no single process that is always reproductive, and because reproduction may occur by a variety of combinations of the many different processes. This leads to occasional confusion, because a single process may not itself be reproductive.

As indicated in a previous paragraph, reproduction itself is impossible to define literally. Any attempt to define reproduction involves all the difficulties of defining individuality, as well as those brought out by the existence of several levels of structure. These levels of structure or individuality are (1) single isolated cells, (2) single cells in a protozoan colony, (3) multicellular animals, (4) fragments of animals no larger than its single cells, (5) composite "animals" of which the members never become complete or isolated, and (6) colonies themselves (producing daughter colonies). (See Figure 2.) These complexities of structure and function add to the difficulties of dealing with reproduction in animals.

It is customary to think of reproduction as increasing the number of individuals, because in the long run this is required for the maintenance of the species. Multiplication of individuals can be accomplished by a single multiple reproduction (where several new ones are formed either from a single parent or by a pair) or else by repetition of a process that produces additional individuals singly.

Some of the processes involved in these reproductions also occur in other circumstances in which one individual gives place to a different one and changes into it (as in autogamy) or two are mutually changed to give rise to one or two new ones (as in conjugation). In these there is no increase in the number of individuals. We find it inappropriate to include the latter processes as reproductive; they are treated below as parareproductive.

Table 1
MULTIPLICATION AND GENOME CHANGE IN REPRODUCTIVE AND RELATED PROCESSES

	Multiplication	Genome change
Bisexual reproduction	Yes	Yes
Pseudogamy	Yes	Yes
Parthenogenesis (meiotic)	Yes	Yes
Parthenogenesis (ameiotic)	Yes	No
Sporogony	Yes	No
Nongametic reproduction (asexual)	Yes	No
Conjugation	No	Yes
Hologamy	No	Yes
Automixis	No	Yes
Nuclear reorganization	No	Yes

There are also a series of processes having close connections to the actually multiplicative processes. These prereproductive processes are often treated with reproduction, because it is only here that they have significance. These are gametogenesis, conjugation, and reproductive behavior. Gametogenesis is not in itself reproductive — it produces no new or even different individuals, only detached cells. It is, of course, a necessary prelude to all bisexual and all unisexual reproduction. Conjugation produces two genetically changed individuals from the initial two, but there is no multiplication. This is prereproductive because it is a prelude to fission of the protozoan individual; it is treated in a later section as a parareproductive process. Reproductive behavior occurs in many animals. It is one of the processes leading to insemination but because this does not automatically lead to fertilization, it is not really part of reproduction per se.

Even omitting behavior, it is clear that some of the processes related to reproduction are multiplicative and some are not. (See Table 1.) We thus arrive at a four-part grouping of all these processes with reference to change of genome:

1. Processes that result in genome change *but are not* multiplicative (conjugation, nuclear reorganization, automixis, hologamy)
2. Processes that result in multiplication *and* involve genome changes (syngamy, meiotic parthenogenesis)
3. Processes that result in multiplication *but no* genome change (ameiotic parthenogenesis, all asexual processes)
4. Processes that contribute to reproduction *but do not* result in genome change *or* multiplication (reproductive behavior and gametogenesis, from the point of view of the parent)

V. REPRODUCTIVE BODIES

It is possible to view all reproduction as occurring by means of fragments (or cells) detached to form the new individual, with either one fragment or two (fused) being involved. These fragments can be called reproductive bodies:

1. Fragments of unicellular body: (a) binary fission, (b) multiple fission
2. Fragments of multicellular body: (a) fragmentation (architomy), (b) budding, (c) strobilation, (d) gemmulation

3. Reproductive cells that do not require fusion: Agametes (diploid) (by multiple fission) and parthenogenetic ova
4. Reproductive bodies that require fusion: (a) haploid gametes (including some by multiple fission in Protozoa), (b) pyloric budding

It must be noted that in such a system the "fragment" may be as much as half of the original multicellular individual or a protozoan reproductive cell. These reproductive bodies are

1. Gametes, haploid unicellular bodies (all spermatozoa, most ova)
2. Gametes, diploid unicellular bodies (some parthenogenetic ova)
3. Agametes, diploid unicellular bodies (most spores)
4. Agametes, haploid unicellular bodies (so-called spores that fuse in pairs) which are herein assumed to be gametes
5. Blastomeres from two or more zygotes, each produced by fusion of gametes, that fuse into an embryo (composite egg, polygenomic)
6. Multicellular bodies such as gemmules (assembled bodies), where the cells come from various tissues
7. Body fragments (by fission or division), either with advance preparation by duplication of organs (paratomy) or simple fragmentation followed by regeneration (architomy)
8. Buds that grow out from the parent and separate
9. Buds that must fuse in pairs (pyloric buds)

Agametes are usually diploid cells comparable to spores of plants (and sometimes called spores when they occur in Protozoa). They are frequently equipped with a resistant covering which enables them to survive environmental conditions which would be lethal to unprotected cells. They can each develop into a new individual by cell division and differentiation. They occur only in Protozoa and Mesozoa. (The only exceptions to diploidy are some spores said to fuse in pairs. We assume that in these there is karyogamy and that they are therefore haploid isogametes.)

As shown in 1. and 2. above, although gametes are usually assumed to be haploid, ova may be of two types genetically; they may be haploid or diploid, but this difference is usually not directly evident. Where there is to be karyogamy, the nucleus must be haploid. This fusion is loosely called amphimixis but must proceed through karyogamy to produce a new cell with a single nucleus which has a double chromosome complement. Where the ova are to develop parthenogenetically, there are three possibilities. First, many species produce diploid eggs, simply omitting the reduction divisions. This is ameiosis, which is essentially asexual. Second, ova that are haploid but do not fuse in pairs may develop parthenogenetically. In this case the activated egg usually undergoes some process to double the chromosome number to restore diploidy. Third, in a few cases of haploid gametes there is no restoration of diploidy and the adult is haploid also. In both cases of haploid gametes, there has been meiosis, and thus they are both meiotic parthenogenesis.

Unfortunately for simple definitions and for the usual generalizations, there are a few animals in which the new individual is produced by the fusion of cells derived from more than two gametes. How the diploid chromosome number is achieved here does not seem to be known.

Multicellular reproductive bodies are occasionally formed in numbers by migration of cells from several tissues to form a body resistant to otherwise lethal environmental conditions. (These are not the same as reduction bodies as that word is used in

Tardigrada, formed one per individual, which later redifferentiates into a functioning individual again.) Different terms have been given in each group where they occur: gemmules, sorites, podocysts, statoblasts, resting buds, and hibernacula.

This view of reproduction by fragments throws a somewhat different light on the customary distinction between sexual and asexual. Some gametes do not require fertilization, and some fragments called agametes, which are not supposed to fuse, actually do so. A large number of animals are produced by processes involving multicellular fragments, but reproduction is not generally looked at from the viewpoint of these fragments.

Chapter 3

SEXUAL vs. ASEXUAL

I. INTRODUCTION

The terms for reproductive processes are defined in several places in this book, but there are several more general terms that need to be understood before we can discuss reproduction effectively. One of the terms, process, is defined in Chapter 1. Individuality, reproduction, and sexuality are discussed and defined in a later chapter, but brief definitions are given here.

An **Individual** is an organized entity with the potential of performing integrated activities necessary to its continued existence. Note, however, that an individual ceases to exist when it separates into several or when it fuses with another. In each of these situations one or several new individuals are formed, with the original one (or ones) no longer existing.

Reproduction is the origination of new organisms from preexisting ones, but specifically it must be multiplicative, not merely the changing of one or two existing individuals into new one(s). It is always a complex of processes at various levels. There must be new and additional individuals formed, not merely changes in one or two.

Sex is the phenomenon of members of a species existing in two forms or structures where each contributes one set of hereditary factors to the next generation. **Sexuality** is the state of organisms exhibiting sex. It is not universal in animals but occurs in some form in most species of metazoans and in many species of protozoans.

II. THE PROCESSES OF REPRODUCTION

Some of the many processes of reproduction have been listed above. They are so diverse that it has always seemed necessary to classify them and to give terms that will identify the natures of those placed in each group. Unfortunately, these groupings and terms have had to be changed from time to time as knowledge increased. The diversity now tabulated has shown all the previous classifications to be again inadequate, and a new scheme has had to be put forward.

Unfortunately again, nearly all recent textbooks still use a system which is inadequate to cover even the processes now known among vertebrates. No simple system will classify the diversity in reproduction as now understood.

Processes have heretofore been grouped or classified under several simple sets of terms: sexual vs. asexual, bisexual vs. unisexual, gametic vs. nongametic, amphimictic vs. apomictic, and meiotic vs. ameiotic. The variety of these systems is apparently due to the fact that none of them are free of ambiguity and exception. It will emphasize the existing diversity to discuss these term-systems and their difficulties.

A. Sexual vs. Asexual

Nearly all zoology textbooks treat reproduction as either sexual or asexual. Most define sexual reproduction as being accomplished by fusion of gametes from two individuals, and they then treat parthenogenesis as an exceptional sexual process. Occasionally "sexual" is so rigidly defined as involving two individuals that parthenogenesis is treated as asexual. In both of these schemes the definitions are inadequate to take account of the variety of processes recognized in the present work, most of which are simply not mentioned in general works.

These same authors generally recognize that in sexual reproduction gametes from two individuals are involved, and their fusion to form a zygote, but they refer only secondarily to the forms that involve only one gamete. They may note that sexual reproduction involves meiosis, but they do not note that meiosis can occur in processes otherwise neither sexual nor reproductive. They usually fail to note that there are other fusions than just the fusion of spermatozoa and ova from two individuals. Fusion of gametes is of course cited, but karyogamy is seldom clearly stated to be the essential act, leaving several activation processes unmentioned.

Problem: Does "sexual" include parthenogenesis? If it does, how about ameiotic parthenogenesis, in which the offspring are genetically identical with the mother and so are exactly comparable to asexually produced offspring? If it does not, how about pseudogamy, in which spermatozoa are present but there is no karyogamy? Either way, the forms of parthenogenesis cause difficulty.

Problem: Does "sexual" include isogamy, in which the individuals apparently are all of one sex? If it does, "sexual" can include unisexual and therefore should include parthenogenesis. If it does not, we admit that syngamy can be asexual, if the gametes happen to be isogametes.

Problem: Does "sexual" include conjugation, in which there are no offspring but there is genetic change in two individuals? If so, sexual need not involve multiplication of individuals. If not, meiosis and exchange between individuals can be asexual.

The word "sexual" is evidently not specific enough for any but the most general use. In ordinary use it is not synonymous with the presence of sex involving male and female individuals. It does not correlate exactly with fusion of ova and spermatozoa, nor with the presence of meiosis. These are some of the reasons why we have been forced to adopt the clumsy expression "Basic Bisexual Reproduction" to specify the real sexual reproduction (as defined later).

B. Bisexual vs. Unisexual

Sometimes a third term is added by subdividing sexual, giving **bisexual, unisexual,** and **asexual.** This takes care of some parthenogenesis, in which the sexual individuals produce only one type of gamete which requires no fusion, but each develops directly into a new individual. However, it does not suffice to classify the processes that are not clear-cut as to sexuality, such as isogamy, hologamy, automixis, and nuclear reorganization.

These terms unisexual and bisexual are sometimes used, but they have very different meanings when applied to species than when applied to individuals. A species is unisexual when its individuals include no males (neuters are usually neglected here). An individual is unisexual when it has only ovaries or only testes, but other individuals of the other sex may occur. A species is bisexual when there are both male and female individuals, but a bisexual individual would be an hermaphrodite. These terms should be avoided unless the application is made clear.

Problem: In some dioecious animals there is no karyogamy. The species are bisexual because there are two types of individuals, but the reproduction is unisexual. Here, the dioecism is irrelevant because there is no cross fertilization.

Problem: A bisexual (dioecious) species has individuals of each sex, but a monoecious species, which has the same two sexes, may reproduce in an exactly comparable manner (cross–fertilizing hermaphrodites). The hermaphrodite is bisexual in a different sense, however, and the gonochorist is unisexual. Is it possible to avoid confusion when we have individuals in a bisexual species with two sexes, one, or none?

Problem: There are bisexual species in which some of the reproduction is unisexual (parthenogenetic or asexual). There are unisexual species (in which only females exist) which cannot reproduce unisexually because the females require activation by males of another species.

Problem: If a given ovum can develop either with or without fertilization, the species obviously cannot be labeled as either bisexual or unisexual, but merely as sexual. The individual is sexual, but its reproduction can be either bisexual or unisexual. (It may, of course, also reproduce asexually.)

There are thus three things referred to by these terms: (1) the species, (2) the individuals, and (3) the reproductive acts. With reference to a single given act of reproduction, it may be bisexual (fusion of gametes), the individuals may be unisexual (gonochorists), and the species may be bisexual, all at the same time.

C. Gametic vs. Nongametic

It has also been attempted to divide all reproduction into **gametic** or **nongametic**. However, both bisexual and parthenogenetic processes use gametes, and there are still the problems of isogamy, hologamy, nuclear reorganization, and automixis. (We cannot here use the otherwise appropriate term **agametic** because it refers directly to agametes.)

Problem: Both bisexual and parthenogenetic processes use gametes, but is reproduction by an unfertilized ovum gametic, whereas reproduction by an agamete is not? The ovum may be either haploid or diploid. According to some writers, agametes also may be haploid or diploid.

Problem: If the use of the term gametic is desired, do "gametes" include isogametes and hologametes as well? How about the fusing structures in autogamy and nuclear reorganization?

Problem: Does the fusion have to result in one individual or can it include a result of two individuals (conjugation)?

These problems have prompted the proposal of the terms "amphimixis" and "apomixis."

D. Amphimictic vs. Apomictic

The mixing of genetic materials through union of gametes from different parents is **amphimixis.** This is usually by anisogametes (spermatozoon and ovum) but may be by isogametes (indistinguishable gametes). This word is usually essentially synonymous with outbreeding and crossfertilization and mixis.

It is probable that most users of the term amphimixis also mean essentially what is herein called "Basic Bisexual Reproduction," because the word involves two organisms and thus fusion of gametes. Amphimixis occurs in all gonochoristic or dioecious animals, when there is karyogamy, as well as in cross–fertilizing hermaphrodites.

Reproduction without karyogamy is **apomixis** — the opposite of amphimixis. Apomixis can be either unisexual or asexual, and in the former either gametic or nongametic. The gametic apomixis includes development of the egg without complete fertilization in which the ova are produced and spermatozoa are either not produced or do not carry through to karyogamy.

Nongametic unisexual includes cytogamy, hologamy, and nuclear reorganization. (This has also been called **agamy.**) Asexual apomixis includes all asexual reproduction in the usual strict sense.

Problem: Amphimixis is fusion of gametes (syngamy) carried through fusion of the nuclei and genomes (karyogamy), but does the word gametes include anisogametes, isogametes, and hologametes?

Problem: Apomixis, the absence of karyogamy, can be sexual (unisexual) or asexual. If sexual, it can be gametic or not (hologamy, cytogamy). How can all of these be covered by a single word?

We are forced to conclude that the basis for these distinctions is not the most basic one available and that these terms can be used effectively only with overt recognition of their

E. Meiotic vs. Ameiotic

It might appear that meiotic and ameiotic could be used to classify the processes.

Problem: Meiotic processes are usually assumed to occur in all reproduction that is sexual in almost any sense, but a large part of the cases of parthenogenesis occur without meiosis, the ova remaining diploid. In many other cases it is not clear whether the ova are diploid or haploid.

Problem: In some cases, haploid ova become "zygotes" (are activated) without influence of male, without becoming diploid, so that the adult animal is haploid and produces gametes without meiosis.

Problem: Ameiosis includes asexual as well as merely those sexual forms lacking meiosis.

In summary, to be sexual the reproduction must involve meiosis and to be bisexual it must involve the fusion of gametes. To be outbreeding the gametes must come from distinct genotypes. To be asexual, there must be no gametes, no fusion, and no meiosis.

F. Meiosis and Genomes

In trying to define sex, sexuality, and sexual reproduction, we have been forced more and more toward a genetic distinction. This has resulted because sexual and nonsexual reproduction cannot always be distinguished by the presence or absence of gonads, or of the production of gametes (either two kinds or one), or of karyogamy, or of any other single feature examined.

It is possible, however, to make a chart of reproductive processes to illustrate the degrees of genetic diversity in the resulting offspring. If mutations are excluded for the moment, new individuals that arise by any process that is not meiotic (or mictic) are theoretically genetic identicals of their parent. Such is the case with animals which are produced unequivocally by asexual means. On the other hand, new individuals arising from processes which include meiosis may show variation in the amount of genetic diversity. Assuming meiosis, the amount of variation will depend largely on two factors: (1) whether the genotype of the offspring was of biparental origin or from a single parent (either parthenogenetic or hermaphroditic), and (2) the relatedness of parent genotypes when biparental. It is during the reduction divisions (or division) of meiosis that such secondary genetic diversity as crossing over between chromosomes occurs.

The interesting possibility of meiosis in obligate parthenogenesis producing no diversity in the offspring occurs when the individuals are all homozygous. Because we know of no instance where this (obligate parthenogenesis with meiosis) occurs, we do not take account of this possibility in this discussion.

Thus, diversity is in general due to three factors: (1) mutation, which occurs in all individuals without effect on sexuality, etc.; (2) chromosome aberrations such as crossing over, in all animals producing gametes meiotically; and (3) fertilization (karyogamy) which brings together and combines two genomes, in the animals usually called sexual (where two gametes fuse).

We conclude that the most basic distinction between sexual and asexual processes is the presence or absence of **meiosis**, which is the principal mechanism that results in genetic diversity. This would produce a clear-cut grouping of reproductive processes as being meiotic or ameiotic (exclusively mitotic). This distinction can be seen in Table 2.

The ameiotic group is clearly not sexual. The meiotic group consists of processes involving either two genomes or only one. These would be called bisexual and unisexual under our meiotic definition of sex. The terms become inappropriate, however, in the case

Table 2
MEIOSIS AND THE SOURCES OF GENETIC DIVERSITY OF NEW INDIVIDUALS

	Mutations	Gamete genomes	Meiosis	Karyogamy
Meiotic processes				
With karyogamy				
Cross fertilization	+	2	2	+
Fraternal fertilization	+	2	2	+
Clonal fertilization	+	2	2	+
Self-fertilization	+	2	2	+
Polar body fertilization	+	2	1	+
Without karyogamy				
Two gametes				
Gonomery	+	2	2	?
Pseudogamy	2	(2)1[a]	(2)1[a]	0
One gamete				
Parthenogenesis (meiotic)	+	1	1	0
Ameiotic processes				
One gamete				
Parthenogenesis (ameiotic)	+	1	0	0
One agamete				
Sporogony	+	1	0	0
Asexual				
(all forms)	+	1	0	0

[a] Signifies that two gametes are present but only one participates in development.

where the gametes are not distinguishable (isogametes) and maleness and femaleness cannot be recognized. It is for this reason that we turn to the meiosis terms as more definite.

In summary, it is possible to classify these processes according to the definitions given. There are four major groups, not two:

1. Sexual reproduction (two gametes and meiosis)
2. Meiotic parthenogenesis (one "gamete" and meiosis)
3. Ameiotic parthenogenesis (one "gamete" and no meiosis)
4. Asexual reproduction (neither gametes nor meiosis)

These distinctions are summarized in Table 3.

III. BISEXUAL REPRODUCTION

Sexual reproduction as usually intended involves two individuals or at least two gonads. This is bisexual. Each gonad produces gametes (gametogenesis) which must fuse in pairs (fertilization) to form a zygote. The gamete fusion (syngamy) must also involve fusion of the gamete nuclei (karyogamy) and thus production of a **synkaryon** or zygote nucleus (a new individual). Each of these requirements is discussed elsewhere.

Many individuals take no part in bisexual reproduction, and it is absent in widely scattered species (i.e., some Hydrozoa, some Polychaeta, and some Rhynchocoela). It is entirely absent from all species in the Chrysomonadina (Protozoa), the Bdelloidea (Rotifera), and the Chaetonotoidea (Gastrotricha).

As suggested above, the two sexes may be one in each individual (dioecism, gonocho-

Table 3
THE REPRODUCTIVE AND GENETIC ASPECTS OF THE PROCESSES

Process	1 No. of individuals	2 Male and female	3a Gamogony	3b Gametes, etc.	4 No. of meioses	5 Karyogamy	6 Multiplication
Bisexual	2	Yes	Yes	Aniso-	2	Yes	Yes
Outbreeding	2	Yes	Yes	Aniso-	2	Yes	Yes
Cional fertilization	2	Yes	Yes	Aniso-	2	Yes	Yes
Self-fertilization	1	(gonads)	Yes	Aniso-	2	Yes	Yes
Parthenogenesis I (pseudogamy)	(2)1[a]	(Yes)	Yes	Yes	(2)1[a]	No	Yes
Parthenogenesis II (meiotic)	1	No	Yes	Ovum	1	No	Yes
Parthenogenesis III (ameiotic)	1	No	Yes	Ovum	0	No	Yes
Parthenogenesis IV (polar body fertilization)	1	No	Yes	Ovum and polar body	1	Yes	Yes
Sporogony	1	No	No	Spores	0	0	Yes
Asexual (fission, budding)	1	No	No	None	0	0	Yes
Conjugation	2	?	No	Nuclei	2	Yes	No
Hologamy	2	?	No	Holo-	2	Yes	No
Automixis	1 or 2	No	No	Nuclei	1	Yes	No

[a] Two individuals but only one participates.

rism) or both in one individual (monoecism, hermaphroditism). The dioecious condition presents no difficulties (except for one mentioned under "Gonad Transplant" in Chapter 2). In monoecious animals there are complications because there often is a theoretical possibility of self-fertilization.

The expression "bisexual reproduction" refers to a process that occurs in many organisms. It can be clearly distinguished in most cases. Unfortunately, the cycles in many of these species also include other reproductive processes, ones that alter the genetic outcome of the cycle. The offspring of these additional reproductive phases, several zygotes or embryos by polyembryony or several larvae by fragmentation, will show less genetic diversity than that in the sexually produced ones.

Thus, in discussing reproduction, it is not enough to define the processes and list their occurrences. It is necessary to show that a given individual, in a certain species, resulting from a given reproduction, may not develop through to adulthood but may obliterate itself through another act of reproduction at some developmental stage. For example, the zygote may undergo polyembryony, whereupon it is transformed into two or more new individuals and thus itself ceases to exist. An individual, therefore, may exist as a zygote for only a brief span until another reproduction destroys it, or it may exist through to the larval stage where fragmentation will destroy it, or it may exist through to adulthood. There is great diversity among animals in this cycle-part which is the life span of a given individual, as shown in Table 2.

IV. BASIC BISEXUAL REPRODUCTION

It is now seen that bisexual reproduction may occur in a cycle which also includes other reproductive processes. These latter may obscure the genetic outcome of the bisexual reproduction. When we have occasion to refer to sexual reproduction that is not obscured by any clonal processes, we need a term to pinpoint reproduction that will or can give the expected genetic ratios in the offspring and to recognize the extent to which this is masked by the occurrence of other processes. We find no term that can effectively be used for this, and so we adopt the phrase "Basic Bisexual Reproduction". This is reproduction in which gametes from two individuals unite in karyogamy to produce one new individual without intervention of any other reproductive processes. Attempts to simplify that into a definition have led to this: **production of a single offspring by immediate biparental karyogamy.** A single offspring must be specified to eliminate polyembryony. The immediacy is necessary to declare that there is no other intervening reproduction, biparental origin is necessary to eliminate self-fertilization and clonal fertilization, and karyogamy is necessary to show that true mixing of genomes occurs.

This Basic Bisexual Reproduction is therefore not a reproductive process but a reproductive cycle consisting of one single specified reproductive process — what might be called "pure" sexual reproduction. It can be looked on as an "uncontaminated" life cycle. The possible contaminants are any additional reproductive processes, such as polyembryony, parthenogenesis, and asexual reproduction, that may occur in the same life cycle.

The term is used here for bisexual reproduction unaltered in its genetic effects by any other reproduction in the cycle or by any "developmental" processes that are actually reproductive. Thus, from each fertilization, it is the production of a single offspring by immediate biparental karyogamy. There is no easy way to indicate where it occurs, except to say that it does *not* occur in any life cycle which includes *any* reproductive process listed in the chart as lacking karyogamy. It is thus much less universal than usually supposed. It can never be assumed unless the entire life cycle is known and no other reproductive processes occur. (See also Chapter 5, "Levels of Diversity.")

This is what is usually understood by cross fertilization, but as previously mentioned there must be additional stipulations as to what is cross and as to the absence of other processes in the cycle. Thus, Basic Bisexual Reproduction involves several restrictions on the usual loose usage of Sexual Reproduction: (1) that it be bisexual, (2) that it not be self-fertilizing, (3) that its results not be altered in that cycle by any apomictic process, and (4) that there actually be karyogamy after meiosis and syngamy.

In determining whether or not Basic Bisexual Reproduction occurs in a given species, it is obviously necessary to know the entire life cycle (and all variation of it) of that species, in order to be sure that the restrictions are met. For this purpose, it seems unlikely that the occurrence of amphimictic reproduction in any species could go undetected. The existence of sex differences is frequently evident externally and is usually demonstrable by dissection. On the other hand, sexual apomictic processes, bisexual and outbreeding but without karyogamy, are often difficult to detect and indeed, in extreme cases, may be proven only by genetic analysis. Sex may be present, even both sexes, without karyogamy occurring, and in many animals even a witnessed insemination will not guarantee karyogamy or absence of apomictic processes (such as parthenogenesis or polyembryony). Furthermore, asexual apomictic reproduction is sometimes not obvious. One cannot anticipate fragmentation, budding, or polyembryony until the process starts.

Chapter 4

THE MANY TERMS AND PROCESSES

I. THE REPRODUCTIVE PROCESSES

Reproduction by means of gametes is **gamogony.** It thus includes all the amphimicitic processes and the parthenogenetic ones (heterogamy and isogamy, amphigony, exogamy, allogamy, dissogeny, endogamy, and automixis) as well as pseudogamy, plasmogony, and gonomery. The production of gametes in any individual is **gametogenesis,** including **spermatogenesis** and **oogenesis.**

The unicellular reproductive bodies produced by gametogenesis are gametes. Their production usually involves meiosis. Thus, such gametes are haploid (possess one genome) but in some parthenogenesis, the ova are mitotically produced and therefore diploid. In most animals the gametes are either ova or spermatozoa, sharply distinct in their structure and behavior — **anisogametes.** This condition is called **anisogamy** of **heterogamy.** In a few Protozoa the gametes are indistinguishable and are called **isogametes,** with the condition called **isogamy.** In most cases of anisogamy, the ovum is much larger than the spermatozoon, and this condition is called **oogamy.** Anisogamy occurs in some Protozoa and in all gamete-producing Metazoa. In some Protozoa the entire (unicellular) animal acts as a gamete (**hologamete**) in a process called **hologamy.**

The words **amphigony** and **amphigenesis** cover all reproduction that requires cooperation of two individuals. They thus include all amphimictic processes, but they may also cover some apomictic processes where both sexes are necessary for activation. **Mixis** is literally the mixing of genetic materials from two individuals, but it usually is assumed to imply that they are unrelated; it is therefore the same as outbreeding. In all of these words, the inadequacy of this assumption of unrelated parents appears when one contrasts inbreeding, which is mixing of genomes from closely related individuals. It is still cross fertilization, still amphimictic, still bisexual, but produces less diversity than outbreeding.

A. Fertilization and Activation

The term **fertilization** is variously used for fusion of gametes, for such fusion followed by fusion of the nuclei, or merely for insemination or even copulation. Because of this looseness in usage, the word is useless as a reproductive process unless it is equivalent with syngamy-plus-karyogamy, which terms are to be preferred. Fertilization covers at least five situations, represented by the terms "cross fertilization," "fraternal fertilization," "clonal fertilization," "self-fertilization," and "automixis" and has been cited also to include somatic fertilization and polar body fertilization. It is always appropriate to use the most explicit term to convey the real nature of the process involved. Even in the case of fusion (syngamy), there are two results which need to be kept separate. Real fertilization brings to the ovum the chromosomes of the spermatozoon, making the fertilized egg diploid. In addition to this, fertilization activates the ovum to develop.

As a reproductive process it covers both the amphimictic karyogamy of Basic Bisexual Reproduction and that of self-fertilization and its relatives. In its broadest sense fertilization cannot really be separated from activation in the cases of parthenogenesis (gametic apomixis), in which there is no fusion.

In ova of parthenogenetic animals, where there will not be any spermatozoan chromosomes added by karyogamy, there still must be some stimulus to activate the "egg" to develop. Activation therefore is not dependent on karyogamy, although it is often a direct result of it.

Problem: Can one use the term fertilization for the fusion of isogametes or of hologametes? How about activation by fusion of an ovum with its own polar body?

Problem: Must the fusing bodies be haploid and must they have been produced by diploid adults?

The term syngamy means literally the fusion of two gametes; some writers also imply karyogamy and others do not. It is also possible to have fusion of the gametes without fusion of the nuclei. The following situations of syngamy occur: (1) cell fusion with the nuclei fusing, (2) cell fusion without the nuclei fusing, (3) cell penetration with the nuclei fusing, and (4) cell penetration without the nuclei fusing. (Activation without syngamy is described in a later paragraph.)

Thus there are at least three activities to be specified in fertilization: (1) penetration or fusion of the gametes, (2) fusion of the nuclei, and (3) activation of the ovum. The second is adequately covered by the term karyogamy — fusion of the gamete nuclei to form a synkaryon. The first is always covered by the term syngamy — the fusion of the two gametes. We use it here in this restricted sense, not implying necessary fusion of the nuclei. The third occurs in all ova that are to survive and develop. It may accompany karyogamy or occur in the absence of karyogamy. We use activation for the latter, because it is only in the absence of karyogamy that we need to refer to it at all. It thus implies only activation in the absence of karyogamy. (See Section I.B.)

Syngamy is thus fusion of two gametes (either anisogametes or isogametes) without specification of what happens to the gamete nuclei. It occurs in most species of most groups of animals, but it is absent in those species using hologamy, conjugation, nuclear reorganization, activation, some cases of parthenogenesis, and exclusively asexual reproduction.

In syngamy, the sources of the gametes (the gonads) can be diverse as regards their distribution among parent individuals. They may be in separate individuals (see Gonochorism and Dioecism, in Chapter 2) or in the same individual. When in separate individuals, there are three situations: (1) reasonably unrelated individuals (cross fertilization); (2) sibling individuals (fraternal fertilization); or (3) clonally related individuals — genetically identical (clonal fertilization). All of these are amphimictic but (2) and (3) produce increasingly less genetic diversity. Although it is implied here that the individuals are gonochorists, it is of course possible that they be hermaphrodites, but the fertilization is between individuals. When the gonads are in the same individual, the syngamy is self-fertilization, which is also amphimictic but produces less genetic diversity. The possibilities of such gamete source are these:

1. In separate individuals (see gonochorism and dioecism, below), (a) reasonably unrelated individuals (cross fertilization), (b) sibling individuals (fraternal fertilization), (c) pairing with gamete from another member of one clone (clonal fertilization)
2. In the same individual, which includes pairing gametes from the one individual (self-fertilization)

Karyogamy is the actual fusion of a spermatozoon nucleus with an ovum nucleus (or the nuclei of two isogametes). This is the essential act of fertilization. It occurs in all fertilizing animals except those listed above under "Syngamy." In less critical work, karyogamic fusion of gametes is more informally called fertilization, as noted above.

Cross fertilization is the situation in which the two gametes come from two individuals that are not part of the same hereditary line. It occurs in some species in many classes of animals but is entirely absent in all species of the order Chrysomonadina of Protozoa, the class Bdelloidea of Rotifera, and the class Chaetonotoidea of Gastrotricha, as well as being absent in occasional species in many other classes of animals. It is also termed **staurogamy.**

As so defined, cross fertilization would be essentially the same as **outbreeding.** (**Exogamy** and **allogamy** are simply syngamy between unrelated gametes and thus the same as outbreeding.) Opposed to this would be **inbreeding,** which is the mating (with karyogamy) of individuals too closely related to be considered "unrelated." Siblings obviously qualify here, but there is no place to draw the line between this and cross- or outbreeding.

A variation on cross fertilization is reported under the term **gonomery** as an apomictic process. The two pronuclei from the fused gametes do not fuse, but remain separate in the zygote and its products. This sounds like parthenogenesis, but because the chromosomes persist we believe they do eventually function in development, making this a case comparable to delayed karyogamy.

Fraternal fertilization occurs when offspring (siblings) of one pair of cross-fertilizing parents produce the gametes which fuse. The resulting zygotes are thus from the combined genomes of the pair of gametes from the two sibling parents, where there is potentially more genetic diversity than in the zygotes from clonal gametes but substantially less than in the case of cross fertilization between unrelated parents.

Clonal fertilization occurs where asexually produced individuals (a clone) give rise to gametes that fuse. Whenever fertilization involves two individuals with basically identical genomes, such as two produced simultaneously by an asexual process from one parent (such as fragmentation), the fertilization is not cross fertilization but similar to self-fertilization in genetic effect. Clones occur in many classes where they frequently go undetected, but clonal fertilization occurs principally in those colonial forms where gametes are produced and fertilized from within the clone, as well as in animals using polyembryony (especially some Insecta).

Self-fertilization (**mychogamy**) occurs in hermaphrodites when an ovum is fertilized by a spermatozoon produced by the same individual. The gametes are thus from a common genotype. It includes autogamy, paedogamy, and cytogamy. It is theoretically possible in most of the classes that include hermaphrodites, even where there is protandry or protogyny.

A similar process in Protozoa that is not actually reproductive is **automixis,** the fusion of sister nuclei. (See Section II.)

Zygotes — The new cell formed by the syngamy and karyogamy is a zygote. It is diploid, formed by fusion of two haploid gametes. It is loosely described as the result of the fertilization of an ovum, implied to be the universal manner of starting a new individual sexually. The fertilization performs two functions: it brings together two genomes to form a new individual, and it activates the new cell to develop.

Unfortunately, many new individuals arise from ova that are not fertilized but merely activated by some other means. The newly activated ovum is not a zygote formed by fusion of two haploid gametes. It has therefore been termed an apozygote — a "zygote" formed without fusion. These occur in all forms of parthenogenesis (see the following section). In the eight forms of activation listed below, only the last produces a zygote. All others produce an apozygote.

1. Activation (parthenogenetic) without influence of males or spermatozoa
2. Presence of males of same species without insemination
3. Presence of males of a related species without insemination
4. Insemination by same species without entry of spermatozoa into ova

5. Insemination by related species without entry of spermatozoa into ova
6. Sperm penetration without karyogamy (gonomery)
7. Fertilization of an ovum by its own polar body
8. Fusion of gamete nuclei (karyogamy)

B. Parthenogenesis

Where an ovum is fertilized (completed karyogamy), it is thereby stimulated to develop with the combined genomes of the two fused gametes giving it the normal double number of chromosomes for that species. Ova that are not fertilized may die, or they may be stimulated in some other way to start development. At some point in this development, if the number of chromosomes is to be restored, there must be some process that is the counterpart of karyogamy.

The process that stimulates the development of the ovum is **activation.** The usual activation is fertilization. The word activation is thus generally restricted to other means of initiating development of the ovum. These include chemical and mechanical influences caused by other individuals (of the same or other species) or by physical forces. Activation includes pseudogamy, plasmogony, and gonomery, as well as the various forms of parthenogenesis. The forms of parthenogenesis correspond to the first six items in the preceding list.

1. In the first instance there is meiosis but only one individual, which is meiotic parthenogenesis, occurring in some Turbellaria, some Annelida, and some Insecta, in which there are diverse means for the restoration of diploidy (zygoidy).
2. There are said to be some species in which males of the species must be present but there is no insemination, the stimulus to develop being apparently a behavioral one. No examples are known to us.
3. In at least one species of fish and one of Amphibia, no males exist, but the presence of a male of a related species is necessary to start development.
4. In some species of Mesozoa, Turbellaria, Nematoda, and a few fish, insemination by males of the same species provides the stimulus with only temporary penetration of the ovum by a spermatozoon.
5. It is reported in some frogs that insemination by males of an unrelated species can provide the stimulus, again without syngamy.
6. There may be fusion of two gametes but no karyogamy. Included here are **plasmogony, plasmogamy,** and **pseudogamy,** all defined as gametic union with the nuclei remaining distinct and thus activation in what is a form of parthenogenesis. (These occur in Protozoa and Nematoda.) Only one term is needed and pseudogamy appears to have been the most consistently used. **Gynogenesis** (I) is the same as pseudogamy (see Glossary.)

Another process usually cited as apomictic is **gonomery,** which is fertilization (or other activation) in which the chromosomes from the fused gametes do not mix but remain in separate groups in the zygote and its products. It seems inescapable that the genes of the male do exert an influence on development.

When there is meiosis and karyogamy, the fusion may be by products of a single cell — sister nuclei, thus from a single genotype, as in polar body fertilization and possibly the form of protozoan self-fertilization called paedogamy. When there is meiosis without karyogamy (only one individual), the genetic diversity is less because only one genome is involved. There is thus no fusion of gametes, and this is meiotic parthenogenesis.

Two peculiar situations have been given names. In **somatic fertilization** spermatozoa are reported to enter somatic cells in the female reproductive tract and exert an influence on them. A similar phenomenon is reported in some Porifera and is reported to affect the reproduction in some way. (This is apparently not a reproductive process, but it involves both sex and cell fusion.) **Merogony** (I, see Glossary) is "fertilization" of a fragment of an ovum that contains no pronucleus. It assumes two forms in Echinodermata: homospermic, where the ovum and spermatozoon are of the same species, and heterospermic, where the two are of different species. **Patrogenesis** is said to refer to development of an enucleated egg induced by fusion with a normal spermatozoon. It is thus scarcely distinct from merogony.

When used to cover all of these, parthenogenesis is still variously defined. In starting the development of the ovum in strict sexual reproduction, the syngamy and karyogamy do two things: they combine two genomes and activate the ovum to develop. In parthenogenesis the egg develops without the second genome, but it must be activated in some way. The act of stimulation is **activation.** Fertilization is the usual means of activation, but in its absence the ovum may be activated by a spermatozoon which does not penetrate it, or which penetrates and then degenerates, or even by the mere presence of a male. The initial influence apparently can be chemical, mechanical, or behavioral.

Other terms have also been used. **Agamy,** meaning without gametes, is sometimes used as a synonym for parthenogenesis although it is a poor word for this, as parthenogenesis always involves one gamete and sometimes involves two (but without karyogamy). It should mean nongametic apomixis, which is herein called asexual. **Amixis,** development without mixing of genetic material, corresponds to activation. It has been defined as the absence of fertilization, which would include all asexual reproduction.

The most general term for activation without fertilization is thus parthenogenesis. It is development of one gamete (always the ovum) without fertilization. (Development from a sperm alone occurs only in plants, where it is called **etheogenesis.**) The ovum (sometimes called a pseudovum) is started on its development by some stimulus other than karyogamy. This stimulus may be penetration by a spermatozoon (without fusion of nuclei), which is pseudogamy or plasmogony. (**Geneagenesis** is the same as parthenogenesis.)

Various forms have received special names:

1. **Arrhenotoky** or **androgenesis,** in which only males are produced (in Insecta)
2. **Thelytoky** or **gynogenesis,** in which only females are produced (only in Insecta)
3. **Deuterotoky** or **amphitoky** or **anthogenesis,** in which both males and females are produced (in Insecta)
4. **Pseudogamy** in which the spermatozoon activates the egg without entering it (in Protozoa and Nematoda)
5. **Plasmogony** or **plasmogamy,** in which the spermatozoon enters the egg but the nuclei remain separate
6. **Gonomery,** in which there is nuclear fusion but the chromosomes remain in separate groups (see also Section I.A. Fertilization)
7. **Progenesis,** which is parthenogenesis in the metacercaria stage of some Trematoda
8. **Polar body fertilization,** in which the second polar body fuses with the ovum and the nuclei fuse in what is called **automictic parthenogenesis** (in various insects)

Several descriptive terms have also been used:

1. Zygoid parthenogenesis is development of an unfertilized ovum that is diploid, either by remaining diploid (ameiotic parthenogenesis) or by restoring its diploidy during development.
2. Hemizygoid parthenogenesis is development from a haploid ovum.

3. Ameiotic parthenogenesis is development from an ovum which is diploid because it undergoes no meiotic division (see Section I. C below).
4. Meiotic parthenogenesis is development from an ovum which has become haploid by a normal process of meiotic reduction divisions.

Meiotic vs. ameiotic parthenogenesis — All of the preceding comments assume that the "ovum" which develops without fertilization is a gamete — that it is produced by oogenesis involving meiosis. Unfortunately, some parthenogenetic "ova" are produced without meiosis, thus remaining diploid throughout. This is ameiotic parthenogenesis, which is said to be even more common than its meiotic counterpart. It is seldom specified in reports of parthenogenesis, so we have usually not distinguished these two in Chapters 6 and 7.

When the parthenogenetic ovum becomes diploid (or if it is already so) and is activated by whatever means, it becomes exactly comparable to a zygote in respect to development. Yet that word is not entirely appropriate, because a zygote is defined as containing a synkaryon, a fusion nucleus. The term apogamete has been proposed for the parthenogenetic ovum, but it seems inappropriate for what is always an (apomictic) ovum. For the activated parthenogenetic ovum we find apozygote to be useful to specify the lack of normal fertilization.

Apozygotes — The activation of an ovum by fertilization (with karyogamy) produces a zygote, the start of a new individual. If the activation occurs in any manner without karyogamy, the resulting initial stage is not quite a zygote, but it is similarly the start of a new individual. It is here called an **apozygote.** (See also the discussion above under "Zygotes.")

C. Asexual Reproduction

The definition of **Asexual reproduction** is unexpectedly simple. It is any multiplication that is entirely independent of sex (of meiosis), but meiosis can only occur in the production of gametes and so it is multiplication without use of gametes. It may occur in animals that have sex or in life cycles which include sexual reproduction, but they are not dependent upon sex nor existence of a second individual. The term thus would include any reproduction not including gametes as well as any parthenogenesis that is ameiotic, but it is the belief of the authors that ameiotic parthenogenesis is sufficiently distinct to be treated as a separate (fourth) class of reproductive processes. All the asexual processes involve isolation of part of the original body to form a new individual, whether it forms by breaking, tissue migration in gemmulation, budding, or production of a unicellular agamete, and whether or not there is complete separation.

There are four basic types of Asexual Reproduction that can be listed. They are (1) production of unicellular reproductive bodies — agametes, (2) production of multicellular reproductive bodies, (3) budding, and (4) fragmentation. Under closer inspection it is evident that the agametes of (1) above are, in reality, of several unrelated types and may involve meiosis, whereas (3) and (4) are indistinguishable in many instances. Furthermore, both budding and fragmentation, although usually described in adults, can occur as early as the embryo, and agametes can be produced either by "adults" or by other agametes.

The term **monogony** is used in Annelida as reproduction which is wholly asexual. It would be more logically used for reproduction involving only one parent. Synonyms for asexual reproduction are **agamogenesis** and **hypogenesis.**

Spores and agametes — The unicellular reproductive bodies called **agametes** occur only in Protozoa, Mesozoa, and possibly the Monoblastozoa (and of course as spores in plants). They take these forms:

1. Amoeboid swarmers produced by multiple fission (Sarcodina)
2. Ciliospores by multiple fission (Ciliata: *Ichthyophthirius*)
3. Flagellated swarmers produced by multiple fission (trypanosomes)
4. Merozoites produced by multiple fission — merogony (Sporozoa)
5. Sporozoites produced by multiple fission (Sporozoa)
6. Spores produced by multiple fission (Sporozoa)
7. Agametes by fission of axial cells (Dicyemida)
8. Agametes by fission of agametes (Dicyemida)
9. Axoblasts in nematogen (Dicyemida)
10. Pseudo-eggs from surface of infusorigen (Dicyemida)
11. Agametes from a plasmodium (Orthonectida)
12. Ciliated swarmers, possibly by multiple fission (Monoblastozoa)
13. Gymnospores from encysted gregarines (Sporozoa) which are true gametes

Agamogony is defined as reproduction by means of agametes, which are diploid reproductive bodies that undergo no union. It is usually called sporogony in Protozoa, but also spore formation or sporulation.

Sporogony — Although definitions differ and usages are at variance, an agamete enclosed in a resistant membrane is usually called a **spore.** One not so protected is more likely to be called a **sporozoite.** Both are produced by sporogony. Sporogony is asexual, but it always follows closely after some process that can be called sexual. It occurs only in the Protozoa (Sarcodina, Sporozoa). **Palintomy** is repeated division, leading to the formation of spores and is thus the same as sporogony. Other terms are sporulation and spore formation.

A curious term is **sporogamy,** "the production of spores by an organism derived from a zygote." It would logically be applied to production of gametes by multiple fission.

In *Gray's Dictionary of the Biological Sciences* under the word "spore," it is remarked that "The meanings of seed, sperm, and spore are very confused." As regards the word spore, "confusion" is scarcely an adequate term. It is described as a caselike structure in which a few sporozoites are formed by multiple fission, as any cell or group of cells capable of producing a new organism, as a product of multiple fission enclosed in resistant covering, as the products of repeated fission of a zygote, as the products of any multiple fission, or as the products of some budding. (There are surely grounds here for confusion!)

Borradaile and Potts as early as 1932 noted that "The term spore is applied to various phases of the life history in a way which is liable to cause confusion." They noted that spores may be gametes, i.e., require fusion in pairs, but they did not clear up the question of what spores are haploid (as these latter must be) and what are diploid (as those from a zygote must be). Any spore which requires fusion (i.e., fertilization) would be difficult to distinguish from a gamete, it would surely be haploid. If some spores are haploid and some diploid, it is certain that some would undergo meiosis, and these would not be asexual.

It appears to us that this must be the result of the confusion brought about by labeling all products of multiple fission as "spores." Some of the products are gametes, haploid, and should be distinguished by that word. All others are diploid, develop directly into new individuals, and should be called agametes (or spores).

Merogony is production of nongametic **merozoites,** which are essentially spores. (It has also been used to describe multiple fission to produce agametes and fertilization of ovum fragments devoid of the nuclei.)

All protozoan agametes are cited as spores in some current works. The agametes of Mesozoa are apparently not clearly different from diploid ova of many parthenogenetic species, but they are not produced in gonads.

Assembled bodies — Multicellular reproductive bodies might be more descriptively termed assembled bodies because their several cells originate from various tissues rather than being formed by proliferation of a single cell type or tissue. Different terms have been given in each group where they occur. The best known are the gemmules of sponges. All of these are formed by the migration of cell types to form a mass which becomes surrounded by a protective covering. It appears that most of these are produced in times of environmental stress and so are reminiscent of reduction bodies, enabling the species to maintain itself through the stress period.

The production of **gemmules** in certain sponges is called **gemmulation** (compare gemmation under parthenogenesis). They are internal aggregations of several types of cells surrounded by a resistant covering, which may be released in some numbers from certain sponges. They may develop directly into adults or first into larvae. (Recent studies make it seem likely that the larvae, at least, develop from zygotes surrounded by nurse cells after fertilization.)

It has been suggested that some gemmules may not serve a reproductive function. Instead of breaking away, they are said to remain *in situ*, and then can re-form the original "parent" sponge after its disintegration.

Production of one gemmule that survives its parent would raise the problem of whether it is a new individual or not; as generally produced in numbers, they are unquestionably reproductive.

There are cell aggregations called **sorites** that are similar to gemmules but without the protective covering. They develop into sponges, but the manner of liberation is not clear. Found only in Porifera.

In sponges, groups of amoebocytes surrounded by epidermal cells are reported to form "reduction bodies" in a degenerating individual; these could develop into new sponges and would thus be multiplicative. They are said to be different from both gemmules and sorites. (The reduction bodies in Tardigrada are merely the regressed body of one individual and so are not reproductive.)

In freshwater bryozoans, the several varieties of **statoblasts** are masses of cells, each surrounded by a chitinous protective bivalve capsule, which can start new colonies. They are produced on the funiculus and consist of epidermal and peritoneal cells and have been described as internal buds.

The term **statocyst** has been cited as a synonym of statoblast. Because it has a standard meaning as a balance organ, this term is best not used where statoblast is intended.

The above are all internally formed structures. There are also two types formed at the surface. **Podocysts** are chitinous cysts in the pedal disk of certain scyphistoma larvae. They break away and develop into ciliated larvae. They appear to be multicellular reproductive bodies rather than fragments. **Hibernacula** are specially modified winter buds (multicellular reproductive bodies) in some Bryozoa. Each consists of a mass of cells surrounded by a thick sclerotized covering.

Fission and division — The terms fission and division have each been applied both to single cells (and Protozoa) and to Metazoa. In spite of the variety of usages, we restrict fission to single cells (binary and multiple fission) and division to multicellular bodies. (The latter can usually be replaced with paratomy and architomy.) They include all asexual processes except perhaps the formation of multicellular reproductive bodies such as gemmules. All of the division and fission processes are included under the little-used term **tomiparity. Fission** is herein restricted to one-celled animals and single cells of Metazoa. The cell undergoes nuclear division, followed by cytoplasmic separation into two new cells.

The splitting of a single cell into two is **binary fission.** It therefore occurs as a reproductive process only in Protozoa and in the Mesozoa where agametes may divide. **Monotomy** has been used with the same meaning.

Any cellular fission in which more than two nuclei are formed followed by separation of the cytoplasm in fragments surrounding each nucleus is **multiple fission.** New cell membranes then form, and more than two new individuals result. This may produce amoeboid young (Sarcodina), flagellated young (Sarcodina, Flagellata), spores (Sarcodina, Sporozoa), merozoites, schizonts, and sporozoites (Sporozoa), gametes (Sporozoa), and agametes (Mesozoa). **Schizogony** is used with various meanings — as multiple fission of a cell, as the production of "several to many buds," as multiple fission that results in new adults instead of gametes or spores, etc. In any case it produces nucleated fragments that eventually become new individuals capable of sexual reproduction, and the fragments (sometimes erroneously called schizonts; see Glossary) do not have a resistant covering such as spores do. (See also merogony, which is largely indistinguishable.) It occurs only in Protozoa (Sporozoa).

Plasmotomy is fission of a multinucleate protozoon (such as an opalinid) into two or more parts without any nuclear division, the nuclei simply being distributed among the daughter cells. In Protozoa without a clearly discernible nucleus, such as *Protamoeba primitiva,* fission is by simple constriction termed **somatic fission.**

Fission in Protozoa may follow a period of growth which forms a new lobe. This lobation has been called budding, but it involves a nuclear division and so is not different from fission. The term budding is therefore not herein applied to these Protozoa.

Division is, in general, the breaking of a multicellular body into two or more (see Fission above). As often used, it is such a general term that it conveys little detail and would therefore be better replaced by more specific terms such as fragmentation, polyembryony, budding, epitoky, etc., even if one did not favor architomy and paratomy. We prefer to keep it as a general term for all these processes but restrict it to reproductive phenomena in Metazoa. The division may occur at any life stage:

1. Zygote or activated ovum (polyembryony) = fission because it is unicellular
2. Embryo, two-cell up to gastrulation (polyembryony)
3. Larva (budding, fragmentation, or strobilation)
4. Adult, including epitoky
5. Bud (secondary budding)
6. Colony, including eudoxy

The common word **fragmentation** is the general term for all division into two or more pieces with or without advance preparation at the site of the break. But when a multicellular animal divides into two, it is usually possible to see either that there has been little or no advance preparation at the breaking point (followed by regeneration of all missing parts on both fragments) or else that all or most of the new organs are formed at the site before the separation. These are called, respectively, architomy and paratomy.

The more restrictive term **architomy** is for fragmentation with little or no prior formation of new organs. The term can be applied to protozoans or metazoans but it is here restricted to the latter. It includes frustulation, pedal laceration, schizometamery, and all multiplicative autotomy. In **frustulation** an irregular piece of polyp stalk (presumably all epidermal) breaks off and forms a nonciliated planula-like larva which will develop into a new polyp (in Hydrozoa only). In **pedal laceration** the polyp creeps slowly over a surface, leaving behind small pieces of the foot, which develop into new polyps (in Hydrozoa, Scyphozoa — scyphistoma larvae, Anthozoa, and the creeping forms of Ctenophora). In **schizometamery,** which is part of epitoky, individual segments are separated out of the middle region of a worm (*Dodecaceria caulleryi*), regenerate both ends, multiply the segments, and then repeat the process, producing in all perhaps 50 individuals.

The term "paratomy" indicates fragmentation or fission in which there has been substantial formation of new organs before the break occurs, the new individual often being completely formed before separation. It includes some forms of epitoky, as well as budding and strobilation.

It must be remembered that in Metazoa the division processes may not be reproductive. **Autotomy** is deliberate breaking off of an appendage or other structure which will presumably then be regenerated. If the broken fragment itself regenerates into a new individual, there has been reproduction. This would have to be specified as "autotomy with regeneration of both parts," which occurs in fairly obvious form in Turbellaria, Rhynchocoela, Asteroidea, and Holothurioidea. It has also been termed **dichotomous autotomy.**

There are two little-used terms for the division of any multicellular organism into two parts, **scissiparity** and **fissiparity.** Neither of them is defined so as to be distinct from either architomy or paratomy. In Annelida scissiparity is used to designate a form of stolonization in which the reproductive stolons become functioning units before separation, which would correctly be paratomy. They are thus not different from fragmentation and are not needed in either general or specific senses. **Schizogenesis** is a little-used general term implying merely fission, either of protozoans or metazoans. It thus includes all asexual processes except perhaps the formation of multicellular reproductive bodies such as gemmules. It is therefore the same as tomiparity.

The group of processes encompassed by the term **polyembryony** amounts to a form of fragmentation that can resemble budding or be merely a separation of cells that have just been produced by cleavage. Thus, division of the zygote or embryo into two or more (up to several thousand) can occur at any stage from the first cleavage up to the gastrula. In most animal groups there is inadequate information on how the embryo (or zygote) divides into several or many. These divisions are all treated here as being the form of fragmentation called polyembryony. Some form of this occurs in Coelenterata, Gymnolaemata, Oligochaeta, Insecta, Mammalia, and very likely others. Something of this sort occurs in Sarcodina, where an hologamous "zygote" divides into several young. The same thing may occur in the unknown part of the cycle of Monoblastozoa. The process of embryonic multiplication in Insecta has been called gemmation, which appears to be an unnecessary synonym of polyembryony.

Identical **twinning** is polyembryony producing two or a few zygote fragments, each of which functions as a new zygote and develops into a complete individual. It occurs in Bryozoa, Insecta, and all classes of Vertebrata and is reported to be sporadic in all major phyla.

In several Trematoda, **successional polyembryony** produces a succession of groups of larvae not derived by metamorphosis or division of a previous stage but by successive fissions of the original zygote and its germinal daughter cells, which have been carried along in each larval stage. Separation of a daughter individual after it has developed into a functioning unit is called **paratomy.** It includes here fissiparity, schizometamery, budding, strobilation, and the separations of epitoky.

The terms "budding" and "fragmentation" (or other forms of division) can usually be distinguished readily, but the difference may be blurred, as in strobilation. In a few cases, both processes may pass under one name, as embryonic budding and embryonic fragmentation both pass as polyembryony. **Budding** is growth of a new multicellular individual from the surface, internal or external, with much diversity in what tissues are involved. Organs in the bud do not always arise from the same tissues as in normal development. The buds may produce more of the same stage (adult from adult) or some other stage (see the material on double budding below).

The word budding is frequently used in Protozoa, where it is said to be endogenous or exogenous. In all cases there is division of the nucleus. We see nothing fundamentally different in this from fission, whereas there is nothing similar to the tissue growth of a metazoan bud. We therefore restrict the term budding to Metazoa and emphasize that protozoan "budding" is merely unequal fission.

There are many descriptive terms for forms of budding. **Colony budding** produces individuals that remain attached to form a colony. In **double budding,** in Hydrozoa, a double bud may be formed, of which the larger part becomes a polyp while the smaller one becomes a medusa. **Dual budding** involves two internal buds that arise in different parts of the body, as in Ascidiacea (see pyloric budding below). **Embryonic budding** is one of the forms of polyembryony in which an early embryo buds off cell masses that become new embryos, as in Calyssozoa. **Endogenous budding** is budding from internal tissues (see especially pyloric Budding). **Exogenous budding** is budding from external tissues, the usual situation. **Larval budding** is budding to produce more larvae, as in Hydrozoa, Scyphozoa, and Ascidiacea. **Multiple budding** can be used to describe simultaneous production of several buds or production of a group of buds from an initial one (as in Ascidiacea). **Pallial budding** is the formation of buds on the mantle in tunicates. It is merely a phrase descriptive of the location of the buds. **Prebuds** are stolon buds from which several generations of specialized buds are formed, as in Tunicata. **Pyloric budding** is the formation of two buds, one from the oesophagus and one from the epicardial tubes, which then fuse to produce one new individual, as in Ascidiacea (see dual budding above). **Resting buds** (in Ascidiacea) are merely buds which after separation develop only after a prolonged interval. They apparently enable the species to survive unfavorable conditions of the environment. **Secondary budding** produces buds upon buds before the first one separates. **Stolon budding** is any budding from a stolon, as in Hydrozoa, Calyssozoa, and Thaliacea. (Budding has also been termed **protogenesis**.)

The word **stolonization** is not much used, but **stolon** has been used in a variety of animal groups for any lateral branch. Such a branch is usually involved in production of new individuals, but it may perform other functions. Some stolons are hollow and connect the body cavities of individuals in a colony. Some apparently are solid. Initially, most of them are involved in producing new members of the colony by budding.

In some books the stolons are considered to be distinct highly modified individuals, but there is usually no direct evidence that they are anything but extensions of the polyps or zooids. They apparently can consist of one type of tissue or of several.

In Polychaeta, the tail region, which bears the reproductive organs, may become separated at sexual maturity and is then called a stolon. In other species, at the tail end clusters of reproductive structures may be formed (by budding) and these are also called stolons individually. This process is often part of the unique reproductive process called epitoky (described below). It is so different from stolon formation in Calyssozoa, for example, that the word conveys little meaning when used in a general sense.

In a few Polychaeta, stolonization into a chain has been called **gemmiparity.** This is strobilation because the terminal stolon is the oldest, and it is paratomy because most parts of the stolon are formed before separation.

Strobilation and chains — Among animals, chains are produced by several processes. First, chains of individuals may be formed by the union of two or more, as in **syzygy**, but this is not reproduction. Chains may be formed by division of an existing body or by budding. The same processes, with rapid separation can result in a series of separate individuals that are not connected after formation. The chains may consist of individuals, stolons, or proglottids.

Chains formed by division into two (repeatedly) are similar in formation, whether the successive parts are individuals, stolons, or proglottids. The term "budding" is sometimes applied to the formation of the new part, but the important feature is usually the growth of a septum between the old and the new parts. An incipient chain consisting of just two parts gives no visible clue to whether it is a strobilus or not. After the chain consists of three parts, it is possible to distinguish between ordinary chains and those in which the septum is always at the end next to the original individual. The latter is **strobilation** which proceeds in such a way that parts are progressively younger from the extremity towards the base. In some chains there is development of new head and tail structures at each septum, as in Turbellaria. In others there are no new structures except the septum itself, as in tapeworms.

Chains that are not formed by strobilation may be found in Ciliata, Turbellaria, Polychaeta, Oligochaeta, and Thaliacea. The reproductive process in these is division, with actual separation indefinitely delayed.

Strobilation produces chains in Flagellata (reproductive bodies from adults), Scyphozoa (ephyrae or scyphistomae from strobilae), Anthozoa (polyp from polyp), Cestoda (multiple gonophores, proglottids, from scolex), Polychaeta (adult or stolon from adult), and Ascidiacea (zooids from buds from zooids).

When a succession of individuals is formed by rapid separation, the process is called **monodisk strobilation.** When the individuals or structures remain attached, it is called **polydisk strobilation.** In Polychaeta strobilation has been called **gemmiparous stolonization.** One of the forms of epitoky is similar to strobilation.

The term **epitoky** is used in quite different meanings for aspects of the life cycle of various polychaetes. In some textbooks and a few invertebrate compendia, it appears that these worms divide into a head end, which is called an atoke and is not reproductive, and a tail end, which contains gametes and is called an epitoke. The epitoke swims to the surface, spawns, and dies. It thus has the nature of a gonophore, a detached fragment that contains the gonads.

In books that treat the subject of epitoky in more detail, it is clear that the word basically denotes the process of metamorphosis (with or without sexual or asexual reproduction or both), in which the rear (sexual) part of the worm changes in appearance but may either detach or remain permanently attached. The diversity of expression of this process is much greater than any of the books make clear.

The textbook usage would be the easiest to describe but would leave a variety of borderline cases that would require other terms or specification. We reluctantly conclude that the second meaning (metamorphosis) is the correct one under the circumstances. It is adopted here with a diagram and a tabulation to show the range of the diversity (see Polychaeta in Chapter 6). (Not all variations are shown, of course, as it is likely that in the Syllidae, particularly, many more combinations and sequences occur.)

Epitoky is thus the process or succession of processes in which one part of the worm body (usually the tail end) changes in structure and physiology in preparation for sexual reproduction. This part may liberate its gametes after breaking away (as a detached gonophore), may remain attached and regress to the original condition after gamete liberation, or may regenerate the missing front end after breaking away and thus become a complete and new separate individual. It merges with ordinary sexual maturation at one point and with alternation of asexual and sexual generations at another and with strobilation at still another.

This process is not only in some instances more of a developmental process than a reproductive one, but it includes in its range not only sexual reproduction but fission, fragmentation, budding, and strobilation. It is at one point paratomous (the fission prepared in advance) and at the next architomous (the fission unheralded by new organs). The division may be single or multiple. The metamorphosed region may be broken off to spawn and die, may remain attached and die with the head end after spawning, or apparently it may remain attached and later regress to the original form until the next breeding season. The isolated fragment may regenerate the missing head and thus become a separate individual for its brief life or it may remain a headless fragment. Stolons (the isolated parts) may be formed in chains by "random" divisions or by strobilation, where all new sections develop at the base of the chain.

In short, the possible ramifications of this process are numerous. It apparently occurs only in Polychaeta. (Diagrams and descriptions of some of the varieties are given under Polychaeta in Chapter 6.)

Three other terms have been used in reference to these processes. **Epigamy** is epitoky in which the entire preexisting atokous individual is modified, the posterior part to form the epitoke. **Schizogamy** is epitoky in which only the posterior end is modified. **Syntomy** is the same as schizogamy.

II. PARAREPRODUCTIVE PROCESSES

One of the difficulties encountered in classifying reproductive processes was the existence in the Protozoa of several processes that seem to involve sexuality of a sort as well as meiosis but are not multiplicative. They are thus involved in reproduction but are not themselves reproductive, being followed by asexual reproduction of some sort. These include conjugation, hologamy, automixis, and nuclear reorganization. Among all the processes that produce new (i.e., different) individuals, there are thus some that are multiplicative and some that merely change the genotype.

```
Multiplicative (genotype change incidental or none)
    Meiotic
        With karyogamy
            Two gametes
                Bisexual
        Without karyogamy
            Two gametes
                Pseudogamy
            One gamete
                Parthenogenesis (meiotic)
    Ameiotic
        One gamete
            Parthenogenesis (ameiotic)
        One agamete
            Sporogony
        Asexual
            Fragmentation, budding, etc.
Genotype change (nonmultiplicative)
    With karyogamy
        Hologametes
            Conjugation
            Hologamy
        Nuclei only
            Automixis
    Without karyogamy
        Nuclei only
            Nuclear reorganization
```

When two individuals fuse permanently with mixing of genetic materials the process is called **hologamy,** with the entire (protozoan) animal acting as a gamete. Temporary fusion is termed **conjugation.** However, whereas there is normally transfer of nuclei and fusion, there are species in which conjugation involves no exchange, this is **cytogamy.** Its function is not clear.

There are Protozoa that undergo changes in the nuclei without gamogony. Fusion of sister nuclei from a previous division is **automixis.** If the sister nuclei are still in one cell, it is **autogamy** (other usages also occur). If they are now in separate cells, it is **paedogamy.** These are not clearly different from **nuclear reorganization,** where the macronucleus may break up and re-fuse (**hemimixis**) or where both nuclei may be reformed from derivatives of the original micronucleus in the cytoplasm of the cell (**endomixis**).

In discussing the parareproductive processes, it is desirable to eliminate clearly some processes that are not even prereproductive. Fusion in pairs is cited in Section I.A., Fertilization, as including some pairing that does not lead directly to reproduction. These include syzygy, fusion of larvae or adults of flatworms, fusion of buds of some Tunicata, somatic fertilization (fusion of spermatozoa with somatic cells), composite zygotes of some Cestoda and Insecta, and fusion of cells in various tissues. None of these are relevant to this discussion.

One other form of nonkaryogamic union of single cells occurs in the Mycetozoa, which are sometimes considered to be sarcodine Protozoa. The individuals fuse without fusion of nuclei, producing a multinucleate plasmodium. This is called **plastogamy,** but it is scarcely reproductive.

Chapter 5

THE LEVELS OF DIVERSITY

I. INTRODUCTION

The diversity of reproductive "processes" that occur among animals, as listed above, is great enough that one might now assume that it represents all the diversity that occurs. Such an assumption would be a great mistake. When these processes are combined into sequences, life cycles, involving several different processes, the complexity is greatly increased. In addition to this, there is even another level of diversity in the variety of life cycles themselves. Altogether these become so complex that it is nearly impossible to tabulate them or give any complete account of their occurrence. These three levels, each cycle, the various cycles of one species, and the diversity of species cycles within a group such as a class or phylum is suggested here in several ways, of which the first is the diversity of processes already described.

In pointing out the diversity in processes and sequences, it is possible to make a false claim of the genetic effects. When one is dealing with the offspring of one generation (or even successive episodes of reproduction by one pair), it is clear that the theoretically expected ratios among the offspring may be upset by any nonbisexual reproduction that occurs. This is why we put so much emphasis on what we call Basic Bisexual Reproduction, which is the only reproduction that routinely yields the expected ratios (Figure 2).

On the other hand, if one is looking at populations, large numbers of individuals produced by many reproductions by many individuals, the introduction of clones among the bisexually produced individuals will quite possibly not produce detectable alterations of the ratio. This is because the clones themselves may be as diverse as the gene pool. It does not follow that there will be perfect correspondence with expected ratios, but only that the upsetting occurrences will tend to balance each other out, and any residual deviation will not be detectable in a large population.

Several words basic to this chapter require explanation or definition. They do not form a single system but are different ways of looking at the reproduction in relation to the life cycle. **Process** has already been defined in this connection as the one act that produces a new individual, not necessarily a new adult nor a similar form. Many, probably most, species employ more than one reproductive process. Reproduction in many animals is a **sequence** of processes, and the sequences are even more diverse than the processes. A sequence is the pathway by which one individual produces a duplicate of itself, frequently after a series of processes. This is the same as cycle, properly **life cycle,** except that many life cycles involve two or more alternative sequences, two or more pathways to reach the final goal of a similar individual.

The sequence of processes is a series of steps, always diverse. Each step is here called a **segment,** a step in the reproductive cycle involving one occurrence of one of the processes. If a pair reproducing bisexually or an individual reproducing either parthenogenetically or asexually survives the process and repeats it several times, each time is an **episode,** the production of one or more new individuals at one time by a given process.

II. SEQUENCES OR CYCLES

Cycles are diverse in several ways; (1) the **processes** involved may differ, (2) the **number of steps** in the cycle ranges from one to at least five in obligate cycles and to some

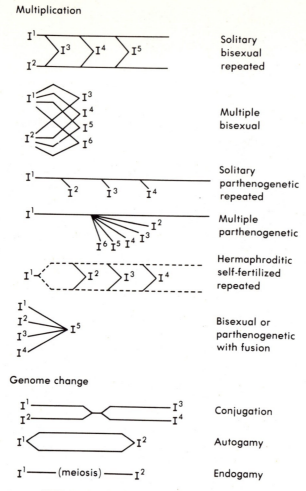

FIGURE 3. Diagrams of reproductive episodes.

larger number in facultative sequences, (3) the **order** in which the processes occur is highly variable, (4) the **number of new individuals** produced by each episode varies, (5) the **number of episodes** of each process varies and is sometimes large, and (6) the various processes occur not only in the adult stage but in all **developmental stages** from the first cleavage of the zygote. All these diversities occur in various combinations, and the resulting life cycles are endlessly diverse (Figure 3).

Processes — The various processes were discussed in previous chapters.

The **number of steps** — In higher vertebrates it is usual for the cycle to include just one reproductive step (segment) repeated in subsequent episodes; but where polyembryony occurs, there is an additional step in the cycle. Again, in *Obelia*, the budding of a hydranth is followed by the budding of a gonangium which is followed by the budding of medusae, before the sexual part of the cycle produces the initial polyp type again. In aphids there will be an indefinite but substantial number of parthenogenetic steps before a bisexual one occurs again.

The life cycle of a species is usually illustrated as a circle: from an adult around the cycle to the adult again. In reality, of course, the circle is a helix or spiral, because the second adult is not the same as the first but merely occupies the same place in the cycle. When a cycle consists of several processes the order in which they occur may vary, this

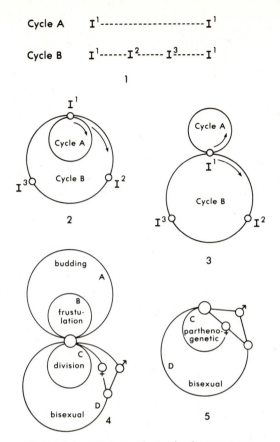

FIGURE 4. Diagrams of reproductive sequences.

being either facultative or obligate. For example, in *Obelia* most individuals in the colony are produced asexually before any sexual reproduction occurs. On the other hand, where there is occasional polyembryony (as in humans) the sequence of sexual reproductions is only occasionally upset by the asexual episode. Whereas most insects reproduce bisexually indefinitely, parthenogenesis is the common method in some, only occasionally interrupted by sexual reproduction. In some polyps, the frequent asexual division is regularly interspersed with a bisexual episode but also occasionally by a different asexual one — pedal laceration or frustulation.

Order in the sequence — There is great diversity in the order in which processes occur in the life cycle of the species. In one life cycle there may be polyembryony early in the cycle and bisexual reproduction at the adult stage. In another there may be first bisexual reproduction in the adult and then asexual in the form of fragmentation of the same individual. There may be parthenogenesis before or after the bisexual reproduction. All of these are normally obligate sequences. There may be a wide variety of facultative sequences where any of several asexual processes and either parthenogenesis or bisexual reproduction *may* occur. Of course, for there to be a specific order in the sequence, there must be more than one process, each giving rise to new individuals.

Number produced — It is hardly necessary to illustrate this. Many mammals produce one new individual per episode; some produce litters of several. Some marine animals produce thousands of ova and thus potentially many offspring from that one episode. Most asexual processes produce only one or two new individuals at a time, whereas the others often produce many.

42   The Diversity of Animal Reproduction

Table 4
SOME OF THE DEVELOPMENTAL STAGES AT WHICH REPRODUCTIVE PROCESSES MAY OCCUR, WHICH THEN BECOME TERMINAL FOR THAT INDIVIDUAL

	Stage	1st reproductive process	Last stage in development	2nd reproductive process	Stage produced	Example
1	zygote	bisex. repro.	adult	fragmentation	new adults	*Lineus*, Rhynchocoela
2	zygote	bisex. repro.	larva	"budding"	new larvae	Cestoda
3	zygote	bisex. repro.	late embryo	multiple budd.	new embryos	Gymnolaemata
4	zygote	bisex. repro.	early embryo	polyembryony	new embryos	parasit. Hymenoptera
5	zygote	bisex. repro.	scyphistoma	strobilation	ephyrae	*Aurelia*, Scyphozoa
6	zygote	bisex. repro.	larva	strobilation	new larvae	Polyclinidae, Ascidiacea
7	zygote	bisex. repro.	adult	parthenogen.	activated egg	Cladocera, Crustacea
8	"zygote"	hologamy	"zygote"	fission	"young"	Lobosa, Sarcodina
9	"zygote"	sporogony	oocyst	mult. fission	sporozoites	Myxosporidia, Sporozoa
10	"zygotes"	fission	"zygote"	autogamy	changed adult	*Actinophrys*, Sarcodina
11	activ. egg	parthenogen.	early embryo	polyembryony	new embryos	poss. in gall wasps, Insecta
12	activ. egg	parthenogen.	larva	fragmentation	new larvae	poss. in Oligochaeta
13	embryo	mult. budding	adult	bisex. repro.	new zygote	Gymnolaemata
14	embryo	polyembryony	adult	fragmentation	new adults	poss. in Oligochaeta
15	larva	budding	larva	bisex. rpro.	zygote	Cestoda
16	larva	fragmentation	larva	parthenogen.	activated egg	poss. in Oligochaeta
17	larva	strobilation	larva	bisex. repro.	zygote	Polyclinidae, Ascidiacea
18	adult	fragmentation	adult	fragmentation	new adults	*Zeppelina*, Polychaeta
19	"adult"	fission	"adult"	conjugation	exconjugants	*Paramecium*, Ciliata
20	"adult"	fission	"adult"	fission	new adults	Protozoa
21	bud	budding	adult	fragmentation	new adults	Hydrozoa
22	bud	budding	adult	frustulation	new larva	Hydrozoa
23	bud	budding	bud	fragmentation	new buds	*Distaplia*, Ascidiacea
24	epitoke	epitoky	epitoke	bisex repro	zygotes	Polychaeta
25	swarmers	mult. fission	"adults"	bisex. repro.	zygotes	Dinoflagellates, Flagellata

Note: Read, for example, line 1 thus: a *zygote* produced by *bisexual reproduction* lives to the *adult* stage where *fragmentation* ends its existence and produces new *adults*, as in some Rhynchocoela. Again in line 5, a *zygote* develops to the *scyphistoma* stage where *strobilation* ends its existence and produces *ephyrae*, as in some Scyphozoa.

The number of episodes — In some asexual reproductions such as fragmentation or fission, the parent may be obliterated in the production of the two or more new individuals. When this does not occur, in budding and in such sexual reproduction as in birds and mammals, it is common but not universal for any particular type of reproductive process to be repeated several times by that pair (or parthenogenetic female) (Figure 4). Thus, there may be successive episodes of sexual reproduction in a pair of adults. There may be repeated strobilation (as in Coelenterata). There may be repeated budding by an adult. These repeated episodes each produce new individuals.

Developmental stage — Nearly all of the individual processes listed in Chapter 4 may occur at other stages than just the adult. For example, budding and fragmentation may occur in the embryo, or sexual reproduction may occur in a larva. (See Table 4.) At least the following occur:

1. Zygote — division at first cleavage (polyembryony)
2. Germ ball — in embryo, successive fragmentations
3. Embryo — from first cleavage to gastrula, budding or fragmentation (polyembryony or colony budding)
4. Larva — fragmentation, budding, or strobilation, and sexual reproduction that may be by either fertilization or parthenogenesis

Table 5
LIFE HISTORY STAGE AT WHICH REPRODUCTION OCCURS IN SELECTED SPECIES

Starting stage	Initiating process	Next reproducing stage
Plasmodium vivax (Protozoa)		
gametocytes (gametes)	fertilization	ookinete
ookinete	sporogony	trophozoites
schizont	merogony	trophozoites or gametes
Obelia sp. (Hydrozoa)		
medusa (gametes)	fertilization	hydranth
hydranth	budding	hydranth
hydranth	budding	gonangium
gonangium	budding	medusa
Lineus sp. (Rhynchocoela)		
adults (gametes)	fertilization	adult
adult	fragmentation	adults
Asplanchnia sp. (Rotifera)		
adults (mictic ova + sperm)	fertilization	adult
adult (mictic ova)	activation	adult
adult (amictic ova)	parthenogenesis	adult
Bugula sp. (Bryozoa)		
autozooid	budding	(avicularia, etc.)
zooids (gametes)	fertilization	ancestrula
ancestrula	budding	autozooid
autozooid	budding	autozooid
Aphis sp. (Insecta: Homoptera)		
adults (gametes)	fertilization	adult
adult (ovum)	activation	adult
Ageniaspis fuscicollis (Insecta: Hymenoptera)		
adults (gametes)	fertilization	adult (possible)
adults (gametes)	fertilization	embryo
embryo	polyembryony	adults
Dasypus novemcinctus (Mammalia)		
adults (gametes)	fertilization	adult (possible)
adults (gametes)	fertilization	zygote
zygote	polyembryony	adults

5. Adult — fragmentation, budding, fertilization, or parthenogenesis
6. Bud — further budding immediately
7. Stolon — budding, strobilation

The reproductive processes of the embryos and larvae (as above) are often followed by a sexual process in the adult, producing a sequence of reproductive processes in what seem to be a single life cycle. For example, a larva (L_1) may reproduce by budding, and each budded larva (L_2) may also reproduce by budding, and each budded larva (L_3) goes on to become adult (A_1) to reproduce sexually, but the first larva (L_1) and the second (L_2) and the third (L_3 which becomes adult) are all different individuals. Actual examples of these short-lived individuals are shown in Table 5.

Table 6
OCCURRENCE OF VARIOUS TYPES OF SEGMENTS IN THE PHYLA AND MAJOR CLASSES

	Dioecious cross-fert.	Hermaphrod. cross-fert.	Hermaphrod.	Self-fertil.	Clonal fert.	Fraternal fertiliz.	Parthenogen.	Agamogony	Adult division	Larval division	Embryonic division	Polyembryony	Parareprod. and others
Protozoa	X					X		X	X	X		X	X
Mesozoa			X			X		X	X			Succ.	X
Monoblastozoa						?			X				
Coelenterata													
Hydrozoa	X		X	X	X	X	?		X	X	X	X	
Scyphozoa	?		X	?		X			X	X			
Anthozoa	X		X	?	X	X			X	X			
Ctenophora			X	?		X			X				
Platyhelminthes													
Turbellaria	X		X	X		X	X		X				
Trematoda	X		X	X		X						Succ.	
Cestoda	X		X	X	X	X	X			X		X	
Cestodaria				?		X							
Rhynchocoela	X	X		?		?	?						
Acanthocephala	X	X				X	?		X				
Rotifera	X						X						
Gastrotricha	X		X	X		X	X						
Kinorhyncha	X					X							
Priapulida	X					X							
Nematoda	X		X	X		X	X						
Gordioidea	X					X							
Calyssozoa			X	?	X	X							
Bryozoa	X		X	X	X	X	X		X		X		
Brachiopoda	X		X	?		X	?		X		X	?	
Phoronida	X		X	?		X			X				
Mollusca													
Amphineura	X		X	X		X	?						
Solenogastres	X		X	?		X	?						
Gastropoda	X		X	X	?	X	X						
Bivalvia	X		X	X		X	?						
Scaphopoda	X					X	?						
Cephalopoda	X		X	?		X	?						

Sipunculoidea							?				×					
Echiuroidea	×					?					×					
Tardigrada	×						?				×					
Annelida																
Polychaeta	×						?	×			×	×				
Oligochaeta		×				×	×				×	?				
Hirudinea		×									×	×				
Archiannelida					×		?				×	?				
Dinophiloidea	×										×	?				
Myzostomida	×			×							×	?				
Pentastomida	×										×	?				
Onychophora	×										×					
Arthropoda																
Merostomata	×										×	×				
Pycnogonida	×										×	×				
Arachnida	×				×						×	×	?	×		
Crustacea	×								×		×	×	?	×	×	
Pauropoda	×										×	×	?	×	?	
Symphyla	×										×	×	?	×	?	
Diplopoda	×										×	×	?	×	?	
Chilopoda	×										×	×	?	×	?	
Insecta	×				×		×	×			×	×	?	×		×
Chaetognatha				×			×				×	×	?			
Pogonophora	×						×				×	×				
Echinodermata																
Crinoidea	×					×					×	×	?			
Asteroidea	×					×	×		×		×	×	×	×		
Ophiuroidea	×					×	×				×	×	?	×		
Echinoidea	×					×	×				×	×	?	×		
Holothurioidea	×						?				×	×	?	×		
Pterobranchia	×					×	?				×	×				
Enteropneusta	×					×	?				×	×				
Tunicata																
Larvacea	×					×	?				×	×	?	×	×	
Ascidiacea	×				×	×	?				×	×	?	×	×	
Thaliacea							?				×	×	?	×		
Cephalochordata	×					×	×				×	×				
Vertebrata	×					×	×				×	×	×	×		

III. REPRODUCTIVE PROCESSES IN A SPECIES

These sometimes occur singly, so that reproduction in a given species may consist of just one such process repeated in each generation, or it may instead consist of several processes each of which results in a new individual (or individuals) and which may occur in sequence in the life cycle or in multiple (parallel) pathways. Each of the individuals will undergo appropriate development, and the species reproduction will be the sum of all the processes producing new individuals among all its members.

Our previous definition of an individual shows that it is that part of the life cycle of that species from the moment of one reproduction to the moment at which that new individual dies or is replaced through fragmentation by several new ones or loses its identity by fusion with another. In the reproduction of one species, the individual is thus one segment in one of the reproductive sequences found in that species. (See Table 6.)

The sequences and segments can be illustrated here. (Note that in all such formulas, the second I_1, for example, is not the same as the first I_1 but merely the next reproductively comparable individual in the cycle.)

1. A sequence: I_1 --- I_2 --- I_3 --- I_1 (as in *Obelia*, where I_1 is a hydranth, I_2 is a gonangium, and I_3 is a medusa)
2. A segment of this sequence: I_1 --- I_2 or I_2 --- I_3 or I_3 --- I_1, each representing one individual. (In Figure 3 in cycles A and B, these individuals are represented by I, regardless of superscript, and the line from one to the next will be one segment in the sequence)
3. The sum of the sequences in a species, as in a scyphozoan medusa (I_1), which may bud or be reproduced via a scyphistoma (I_2), or in which the scyphistoma may bud more of itself (I_2a)

$$I_1 --- I_1$$
$$I_1 --- I_2 --- I_1$$
$$I_1 --- I_2 --- I_2a --- I_1$$

These sequences and segments may be listed in descending order thus:

1. All the reproductive sequences of a species (one or more)
2. A single sequence from among those in a species (consisting of one or more segments)
3. A single segment in any sequence in the species (between any two successive individuals) (see Table 7)

Individuals in a succession of segments frequently succeed each other in either fixed or variable sequences, producing what is often called **metagenesis,** or alternation of generations. The alternations may involve bisexuality, parthenogenesis, and asexual reproduction, each in several forms, and we have tabulated at least 50 basic alternation sequences that occur. Only a few of them can be shown here.

A. Single Segment Sequences

These are the simple cycles that are often assumed to be universal or nearly so. They will produce populations of predictable genetic make-up. There are three basic ones and several varieties.

1. Basic Bisexual Reproduction   S – S – S – S
2. Parthenogenesis   P – P – P – P
3. Asexual reproduction   A – A – A – A

Table 7
THE VARYING SPAN OF INDIVIDUALS BETWEEN REPRODUCTIVE OCCURRENCES

	First individual	End of life stage	Because of this process	Producing next new individual	Example (occurs in)
1)	zygote	early embryo	polyembryony	new embryos	Armadillo
2)	zygote	late embryo	multiple budding	new embryos	Phylactolaemata
3)	zygote	larva	fragmentation	larvae	Scyphozoa
4)	zygote	adult	any asex. repro.	fragments	Polychaeta
5)	activated ovum	early embryo	polyembryony	new embryos	Insecta
6)	activated ovum	larva	fragmentation	fragments	Gymnolaemata
7)	embryo	embryo	fragmentation	fragments	Insecta
8)	larva	larva	fragmentation	larvae	Ascidiacea
9)	adult	adult	fragmentation	fragments	Rhynchocoela
10)	adult	adult	conjugation	exconjugant	*Paramecium*
11)	adult	adult	autogamy	differ. adult	Protozoa
12)	adult	adult	fission	adults	Sarcodina
13)	bud	adult	any asex. repro.	fragments	Anthozoa
14)	fragment	adult	any asex. repro.	fragments	Polychaeta
15)	adult	(irrelevant)	gamet. and fertil.	zygote	Humans

Note: Read line 1 (for example) thus: a new *zygote* ends its life at the *early embryo* stage because of *polyembryony* which produces new embryos, as occurs in the *armadillo*. Again in line 9, an *adult* ends its life at the adult stage because of *fragmentation* which produces *fragments* that regenerate into adults, as occurs in some *Rhynchocoela*.

Each formula represents four generations, passing from left to right. Each letter represents a new individual produced by a process denoted by that letter (S, bisexual; P, parthenogenetic; A, asexual). Between (1) and (2) there are actually three other varieties: obligately self-fertilizing hermaphrodites (SF – SF – SF – SF), fraternal fertilization (FF – FF – FF – FF), and clonal fertilization (CF – CF – CF – CF). In addition to this the parthenogenesis can be of either of two sorts, meiotic or ameiotic, and the asexual reproduction can be of any of at least four processes: budding, stolonization, agamogony, or fragmentation.

A further diversity occurs in the sex arrangement of the individuals in species in which Basic Bisexual Reproduction occurs. The species may be dioecious or monoecious. The hermaphrodites would then necessarily be obligately cross fertilizing (in order for the reproduction to be considered to be Basic Bisexual). (Remember that to be a single-segment sequence these must be species in which no other reproduction ever occurs.)

Among the A diagrams of Figure 3, there could be normal bisexual reproduction or parthenogenesis or any of a variety of asexual processes. The most common process is the bisexual, and it has been our major purpose to isolate this from all other reproductive processes.

The only reproduction which will produce the Mendelian ratios is the one here called Basic Bisexual Reproduction, defined above as production of a single offspring by immediate biparental karyogamy. *Any* deviation from this or addition to it may alter the gene ratios to a greater or lesser extent.

To determine whether a reproductive sequence qualifies as Basic Bisexual, three things are necessary: (1) there must be only the one reproductive *sequence* employed by that species, rather than several alternate sequences; (2) there must be only one reproductive *process* in the one sequence; and (3) that reproductive *process* must be fusion of gametes with karyogamy.

For example, a species of *Peripatus* (Onychophora) reproduces solely by one sequence, consisting of just one process, which is fusion of gametes with karyogamy. This is therefore Basic Bisexual Reproduction. In *Homo sapiens* this same situation is the usual one, but the fact that humans can and occasionally do reproduce by polyembryony makes this species one with two sequences, in which one sequence consists of just one reproductive process and the other one consists of two. In *Hydra*, the animals may reproduce bisexually (a sequence of one process) or asexually (also a sequence of one). The species thus has two sequences each of one process.

Thus, we are contrasting **Basic Bisexual Reproduction** (as a cycle of just one particular process — karyogamy) with **All Other Reproduction** (nonkaryogamic, mixtures of processes, sequences of processes, mixtures of sequences, and any other single reproductive process).

Lest it appear that reproductive sequences are all much alike, it should be noted that there are more than 200 different sequences; these can be reduced to about 50 basic types. We are attempting to show that the one of most concern to biologists is by no means universal, nor can it be taken for granted in any animal.

B. Multiple Segment Sequences

Sequences involving more than one reproductive process may be of three sorts: (1) they may be obligate sequences, where two processes alternate (in some fixed series); (2) they may be obligate and fixed as processes in the sequence but variable as to the number of occurrences of one of the processes in the sequence; and (3) they may involve a fixed basic sequence with the possibility of occasional interruptions by some other process. These are

1. Completely fixed sequence   S – A – S – A
2. Variable number in an obligate sequence   S – P – P_n – S
3. Occasional alternatives   S – S – S – S
                                                                                    |
                                                                                     A – S
                                                                                    |
                                                                                     A

In the fixed sequences, besides the one shown, there may be P – A – P – A. It is not unlikely that there are also S – P – S – P, although none has come to our attention.

In the variable sequences, P_n signifies an indefinite number of parthenogenetic generations. Besides the one shown, these can involve S and P, P and A, or S and A, and in the latter may involve also a sequence of two different A's, as in a hydroid which reproduces sexually as well as by budding and occasionally by pedal laceration.

The occasional alternations are numerous and diverse. In a sexual series an individual may also reproduce by self-fertilization or by clonal or fraternal fertilization. It may occasionally reproduce by asexual reproduction as shown in 3. or by parthenogenesis. There may be individuals which can reproduce either bisexually or parthenogenetically or asexually, or all three.

The diagrams of Figure 4 show how many species can complete this reproductive sequence, from I_1 around to I_1 again by more than one route. These are shown as the A, B, C, and D pathways in the diagrams, and it is possible that there may be more than four in some species. We call these alternative sequences because they are facultative. They do not occur in a fixed sequence. Depending on how they are drawn, the diagrams will be parallel (linear) or tangential.

By definition, a reproductive sequence of a species is the series of stages (or processes) necessary to produce an individual identical in form to the initial one.

To recapitulate, in the three cases diagrammed above, the segments, the sequence of segments within a species cycle, and the sum of all the sequences that the species may use are as follows:

1. The segment occupied by one individual, from one reproduction to the next. In Figure 3, the individual segments are in A, $I_1 - I_1$, in which they are the same as reproductive sequences (one segment each), and in B, $I_1 - I_2$, $I_2 - I_3$, and $I_3 - I_1$, in which there are three individual segments in the one reproductive sequence. (See Table 7.)
2. Single reproductive sequence of a species. One or more individuals in a single sequence: from $I_1 - I_1$ in all six cycles, but in the B cycles there are several steps involved in each reproductive sequence.
3. Sum of the reproductive sequences of a species. Each species has at least one pathway, but many have alternative pathways, as shown in each pair of diagrams; cycles A and B together in each case; (diagrams could be drawn for other species that would show more than two alternating cycles) (See Figure 3.)
4. Episodes are the repetitions of a single-process sequence by one individual (parthenogenetic or asexual) or two (bisexual), as in all animals that live long enough to take part in two or more breedings.

IV. DIVERSITY WITHIN A CLASS

It is reasonable to assume, when one knows of a certain reproductive strategy in the members of a group of animals, that other members will show the same strategy. This seems reasonable because within the (taxonomic) group the structures and general behavior are usually seen to be uniform, as indeed is normally required for them to be placed together in a group. It is, in fact, always a surprise to find exceptions to the general pattern.

Yet exceptional processes and cycles of reproduction occur with unexpected frequency throughout the animal kingdom, so much so that the only rule must be that diversity is likely. Obligate parthenogenesis occurs in most phyla and polyembryony is either obligate or facultative in many of the classes, including the higher vertebrates. The parthenogenesis is without meiosis in at least five classes. It is extremely common in all parts of the kingdom for a "life cycle" to include successively two or more individuals using different reproductive methods. This also may be obligate or facultative.

It is difficult to illustrate or even to tabulate the diversities at this level. By outlining on one diagram all the methods found in a class, one can get an idea of the diversity within that class, but diagrams do not show all the diversity because the cycles have to be simplified.

On the following pages are shown eight examples of such diagrams, some simple and some complex. Some further complexities have had to be omitted in these two-dimensional presentations, especially where two or more individual cycles come together to form one new individual.

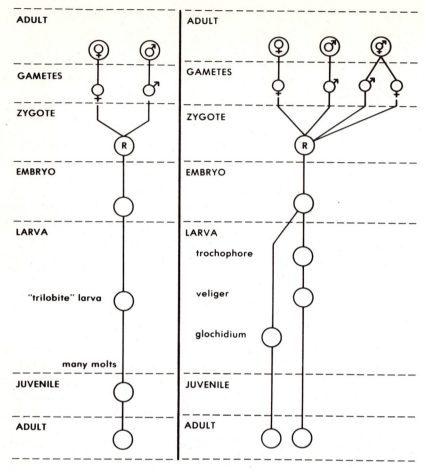

FIGURES 5 and 6. Composite diagrams of the reproductive pathways of the species in the Merostomata and the Bivalvia. In these diagrams the large circles are individuals, sometimes with sex shown. Gametes are shown by sex symbols not in circles. R indicates that there has just been reproduction of some sort and this is a new individual. An apozygote is indicated by apo-.

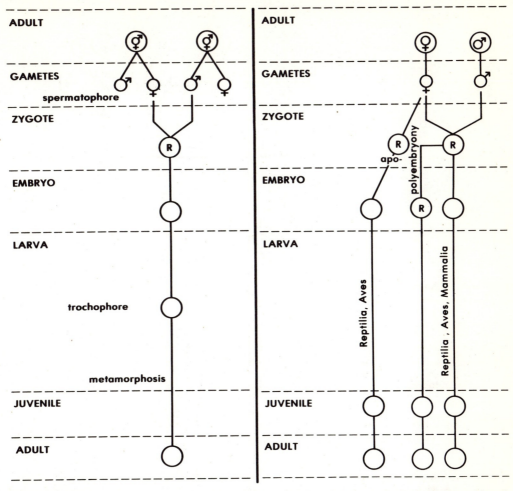

FIGURES 7 and 8. Composite diagrams of the reproductive pathways of the species in the Myzostomida and the higher Vertebrata.

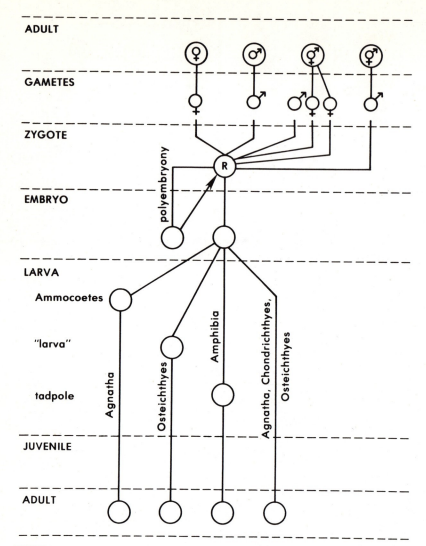

FIGURE 9. Composite diagram of the reproductive pathways of the species in the lower Vertebrata.

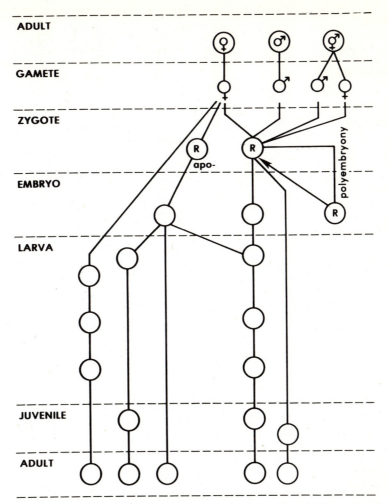

FIGURE 10. Composite diagram of the reproductive pathways in the species of Insecta.

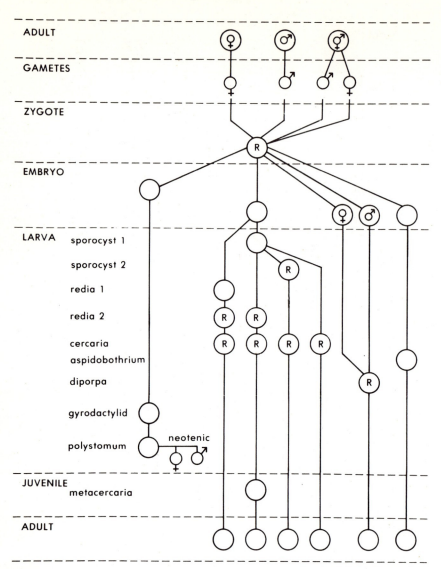

FIGURE 11. Composite diagram of the reproductive pathways in the species of Trematoda.

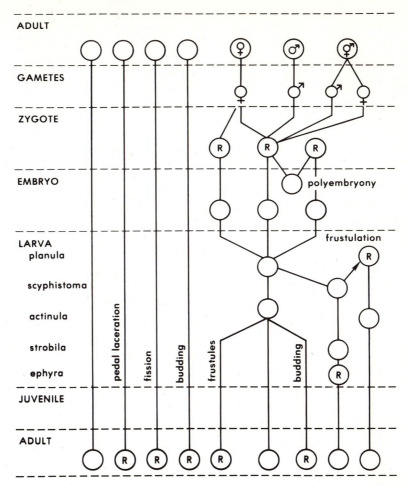

FIGURE 12. Composite diagram of the reproductive pathways in the species of Hydrozoa.

Chapter 6

REPRODUCTION IN ANIMALS BY CLASS

I. INTRODUCTION

This is the definitive listing of the reproductive diversity for every class of animals. A few groups are listed as classes or phyla although they may not deserve that rank; this placement is not intended as final opinion on their status. They are included to permit assessment of their true position.

Certain other information is given to improve understanding of the reproductive processes:

1. Phyla: For each phylum are briefly outlined the **Habits,** to recall the nature of the animal, and the **Reproduction,** as a listing of the processes.
2. Classes: For each class (and each one-class phylum) are cited the following:

Reproductive cycle — A composite description, briefly described examples (often omitted), and reproductive processes including occurrence and circumstances; in this sequence: Basic Bisexual, other sexual, other gametic, and asexual.

Where it is deemed appropriate to cite the source of the particular statement, it is immediately followed by an asterisk; this signifies that the source is listed in the section on Classification and Notes, and the bibliographic citation is then given in the list of References.

II. PHYLUM PROTOZOA

Habits — These unicellular organisms are extremely diverse. They are marine, fresh-water, saprophytic, saprozoic, ectoparasitic, or endoparasitic, and they occur in water films away from bodies of water. They are free-swimming or creeping, attached, or burrowing. They are frequently colonial. (The terms used by various authors make it difficult to describe the processes in uniform fashion.)

Reproduction — Bisexual processes (gamogony and hologamy) occur but apparently are always part of apomictic cycles, so there is no Basic Bisexual Reproduction. Parthenogenesis (plasmogony). Self-fertilization (autogamy). Para-reproduction (conjugation, nuclear reorganization — endomixis and hemimixis, somatic fission). Asexual (binary fission, multiple fission, sporogony or sporulation, plasmotomy, budding, division or syntomy, strobilation, agamogony, polyembryony). Colonial (duplication by constriction and budding of daughter colonies, and by a unicellular agamete).

A. Class Sarcodina

Reproductive cycle — Reproduction is basically by fission. In parasitic forms this may be multiple fission. Temporary fusion of two individuals, plasmogony or conjugation, is often cited as a sexual process. It never results in an increase in individuals, and in our opinion it can be treated as sexual reproduction only if the following separation of the two is considered to be analogous to the first cleavage of a zygote, in which case one would have hologamy followed by polyembryony.

In many Rhizopoda and others there is no sexual reproduction of any sort; the asexual reproduction is by binary or multiple fission, budding, or plasmotomy.*

In *Protamoeba primitiva* there is no clearly visible nucleus and division is by mere constriction.* Here this is called somatic division.

In *Endamoeba histolytica* binary fission is followed by encystment, with the nucleus duplicating twice. The quadrinucleate amoeba then excysts and divides into eight uninucleate amoebulae. These develop into adult amoebae.*

Lobosa reproduce by binary fission, multiple fission, or production, after many nuclear divisions, of numerous internal uninucleate cysts (or spores), which hatch into minute amoebae.* An unusual sexual process is known in *Sappinia diploidea,* where two nuclei originate from the cytoplasmic union of two individuals in a common cyst;* these nuclei fuse before the next fission; this is thus endogamy. In parasitic forms there have been reported amoeboid anisogametes that united in pairs (syngamy). Hologamy is widespread and may result in a zygote, which generally divides into several young. In *Trichosphaerium* adults produce amoeboid agametes by multiple fission. These develop a gelatinous test and sporulate into biflagellate isogametes which unite in pairs and develop into the adult.* Amoeboid anisogametes may be produced to unite in pairs.*

Foraminifera all reproduce by multiple fission, but both agametes and gametes are produced by this means. In *Elphidium* the adults are said to be dimorphic, some being schizonts or agamonts, which will reproduce asexually, the others being gamonts which will produce gametes.* In *Patellina* the diploid schizonts become multinucleate adults. They then undergo multiple fission into uninucleate amoebulae. These develop into haploid gamonts, undergo a complex nuclear reorganization, and then sporulate into minute amoeboid isogametes. These fuse in pairs to form a zygote that develops into a schizont. (In other genera the gametes may be biflagellate.)*

In Heliozoa binary fission, plasmotomy, hologamy, and autogamy occur.

In Radiolaria binary fission and sporulation into biflagellate isogametes occur, but no complete cycle is known. Colonies may be formed.

Basic Bisexual Reproduction probably does not occur. Hologamy, anisogamy, and isogamy occur but always with asexual processes. There is none in Foraminifera (because of multiple fission), and possibly also none in the Radiolaria. A possibly sexual process occurs in *Sappinia diploidea,* where two nuclei originate from the cytoplasmic union of two individuals inclosed in a common cyst, these fusing at the next encystment.*

Plasmogony or temporary cytoplasmic union also occurs in the Lobosa.*

Autogamy in *Actinophrys* and *Actinosphaerium* consists of division into two daughter cells which, after two typical maturation divisions, fuse to form a zygote.* This is also found in the Lobosa.*

Binary fission predominates* and seems to be always mitotic. It is called monotomy by some.*

Multiple fission after nuclear divisions may produce numerous amoeboid young *(Arcella),* * sometimes within a cellulose cyst *(Nuclearia, Vampyrella)* * or producing flagellated swarmers *(Paramoeba).* * Multiple fission occurs throughout the Foraminifera, involving an alternation with isogamy.* (See also Plasmotomy)

Sporogony follows gamogony and results in minute bodies encased in a resistant covering.

Multinucleate individuals may divide by **plasmotomy,** without nuclear division.*

Budding may be either single or multiple;* it may involve the budding of amoeboid or biflagellate young *(Acanthocystis).*

Polyembryony or something akin to this occurs in Lobosa, when hologamy results in a zygote which divides into several "young."*

Colony duplication occurs by constriction, as in Radiolaria, etc.*

B. Class Flagellata

Reproductive cycle—Reproduction is primarily by fission, transverse in the Dinoflagellata, and longitudinal in the rest. Multiple fission or sporulation occurs in the Dinoflagellata. Hologamy is reported but the cycle is unknown. In the Phytomonadina, besides fission, there is sexual reproduction with all grades of isogametes, anisogametes, and hologametes. There is a zygote which undergoes reduction in its first divisions. Colonies are formed by repeated fissions.

No real bisexual reproduction occurs, and asexual processes include longitudinal fission, multiple fission, and budding. Sexual processes include only hologamy, but the reproduction is exclusively asexual.

Basic Bisexual Reproduction apparently does not occur. Isogamy and anisogamy occur in the Phytomonadina, and hologamy occurs in one or two other species.*

Binary fission is generally longitudinal, but it is transverse in the Dinoflagellata. It frequently results in colony formation, and, as in *Volvox,* may form a miniature colony inside the main colony.* It may occur within a cyst in the Dinoflagellata.*

Multiple fission in trypanosomes may produce a number of daughter individuals arranged in a rosette.* In Dinoflagellata there is multiple fission within a cyst, producing flagellated swarmers.*

Although **sporulation** has been reported,* there appears to be nothing that is clearly different from multiple fission.

Budding also has been cited, but it usually appears not to be different from sporulation and multiple fission, except in *Noctiluca.** In *Noctiluca,* a bud forms and then divides into several, as many as 500.*

Fission is a process somewhat distinct from binary fission because most parts are duplicated before separation. This is syntomy which occurs in *Trypanosoma lewisi.**

Strobilation apparently occurs in the dinoflagellate *Haplozoon,* where a chain of cells is formed with the hindermost serially freed as reproductive bodies of some unknown sort.*

Colony division—Colonies may be duplicated by repeated fission of one cell into a daughter colony which escapes from the parent colony by rupture, as in *Volvox.** An entire colony of Chrysomonadina may split into two.* A new colony, in *Pandorina* or *Eudorina,* may also be formed by detachment of a single cell which then produces a colony by repeated fission.*

C. Group Opalinida

The opalinids have recently been transferred from the Ciliata (subclass Protociliata) to the Flagellata (superclass Opalinida). There is still question as to the proper place for this divergent group, so here it is kept separate for convenience.

Reproductive cycle—Anisogamous gametes are produced. The zygote often encysts or develops at once into an adult. There is said to be plasmotomy, but conjugation does not occur.

Basic Bisexual Reproduction by gametes may occur,* but it is generally obscured by fission.

Binary fission is reported, always longitudinal.

Plasmotomy (multiple fission)—The multinucleate opalinid divides by plasmotomy into small cells which encyst, and after encystment they then divide into anisogamous gametes.* This has been termed syntomy.*

D. Class Ciliata

Reproductive cycle—All reproduction is asexual, although the sexual process of conjugation also occurs. Often in a cyst, the ciliate divides by binary fission, either transversely or longitudinally. Conjugation with exchange of nuclei is followed by fission, four individuals resulting from the union of two. This can be looked upon as hologamy, with the exconjugants being the result of a first cleavage. (The Protociliata here are treated separately, as Opalinida.) In the Holotricha the transverse fission may produce chains by incomplete separation of the posterior fragments. In the Chonotricha the fission is replaced by lateral budding.

Basic Bisexual Reproduction does not occur.*

Autogamy occurs, as in *Paramecium,* when the micronuclei divide mitotically to form eight or more, of which two fuse to form a synkaryon.*

Conjugation is the only sexual process;* it does not by itself constitute reproduction.

Nuclear re-organization is common and takes several forms: (1) Endomixis, in which conjugation stops short of nuclear fusion;* and (2) Hemimixis, in which the macronucleus degenerates.*

Binary fission is generally transverse, but it is longitudinal in the Peritricha.* The micronuclei divide mitotically, but the macronucleus divides by a sort of cryptic mitosis* sometimes labelled amitosis.* Longitudinal fission is the principal means of colony formation.* Fission generally occurs in cysts.* Fission may be by paratomy, as in *Stylonychia,* in which the organelles of the daughter organism are formed before separation.*

Budding—Buds are formed laterally in the Chonotricha.* They may be apical, in Astomata, resulting in a chain of individuals.* This is not clearly distinct from fission; it may be strobilation.*

Colony duplication—Apparently colonies are produced by fission rather than by budding. In such a case as *Vorticella,* colony formation is by budding.*

E. Class Suctoria

Reproductive cycle—Although conjugation may precede separation, all reproduction is asexual, usually described as exogenous or endogenous budding. In the latter, the ejected protoplasm forms an "embryo" with cilia. Nonciliated vermiform buds may also be given off; however, the separation is sometimes effectively equal division, with regeneration of each part.

Basic Bisexual Reproduction is not known to occur.*

Fission may occur, with the distal half of the stalked protozoan absorbing its tentacles, putting out cilia, and breaking off to swim away.*

Conjugation is said to result sometimes in complete and permanent fusion of the two conjugants, as in *Lernaeophrya capitata.*

Budding may be exogenous or "endogenous" (in external pouches), with the nuclei behaving as in ordinary binary fission.* It may result in ciliated "embryos," in a single ciliated animal resulting from separation of the distal half, or in vermiform bodies.* Vermiculate branchings appear in *Dendrosomides,* devoid of tentacles, in what may be a form of budding.*

F. Class Sporozoa

Reproductive cycle—Both sexual and asexual reproduction involve multiple fission, into agametes or gametes (either iso- or anisogametes). The gregarines and coccidians differ from all animals in having the reduction of the chromosomes at the first division of the zygote, which is thus the only diploid part of the cycle. Examples include:

Gregarinida—Sporogony produces eight sporozoites, which develop into trophozoites. These may adhere in chains (syzygy), in which the anterior is apparently female, the posterior male. When the trophozoites encyst in pairs, they become gamonts and produce gametes of one sex, usually isogametes, those of each pair fusing to form zygotes which transform into "spores." Inside, sporogony produces sporozoites.

Haemosporidia—In *Plasmodium* the amoeboid sporozoites in red blood cells (where they are called trophozoites) undergo multiple fission to produce merozoites, which may repeat this cycle indefinitely to produce more and more merozoites. Then each merozoite develops into a gametocyte, either male or female. In the next host, the female gametocyte becomes an "ovum," the male gametocyte divides into four or eight "spermatozoa." The zygote or ookinete becomes an oocyst, in which multiple fission (sporogony) produces many sporozoites.

Myxosporidia—The spore hatches into a uninucleate amoeboid zygote. It is a zygote because it arose in the spore from the fusion of two sporoplasmic nuclei that function as gametes. The zygote nucleus undergoes multiple fission to form a plasmodium, after which plasmotomy produces many internal spores by the clustering of six nuclei, two of which become the gametic nuclei which later fuse. This fertilization is thus paedogamous (and isogamous) autogamy.

Basic Bisexual Reproduction does not occur. Sexual processes occur but are masked by sporogony. The cycle is sometimes referred to as paedogamous autogamy, therefore being apomictic.*

Autogamy occurs in Cnidosporidia and is paedogamous.*

Multiple fission is extremely varied, producing gametes by anisogamy or isogamy, merozoites by merogony from merozoites, or merozoites by schizogony from trophozoites. The fission normally follows mitotic divisions of the original nucleus to form a multinucleate body. The zygote formed by gamogony and fertilization also usually undergoes sporogony into naked infective sporozoites or into spores (in which more sporogony may occur).*

Agamogony occurs through multiple fission with the production of agametes (sometimes termed merozoites).*

Plasmotomy—Certain spores of the Myxosporidia hatch into uninucleate amoeboid young (actually a zygote) that becomes multinucleate. If this individual is considered basically multinucleate, the resulting division is simply multiple fission; if it is considered to be basically uninucleate, the division is plasmotomy, as the cytoplasm divides around each nucleus.*

Budding is reported to occur endogenously when a nucleus is cut off with a bit of cytoplasm in the interior of the sporozoan; this then divides in one of a variety of ways to produce anisogametes, in what is essentially autogamy.*

Fusion of adults (permanently) occurs in Gregarinida, in syzygy. The mature gamonts may be visibly distinguishable as male and female; they encyst in pairs, with their gametes (isogamous or anisogamous) of the same sex as the gamont; the gametes unite in pairs to form zygotes called spores.*

III. PHYLUM PORIFERA

Habits—All are marine, except the family Spongillidae (Demospongia), members of which occur in fresh water. All are sessile, living attached to objects or rarely unattached and rolling about on the bottom. Sponges are usually multiple structures that may be looked upon as colonies, but individuality of the parts is impossible to establish. The following descriptions refer to the entire structure. It is reported that (unidentified) sponge larvae or postlarvae may fuse, forming heterogenomic individuals.*

Reproduction—Basic Bisexual. Parthenogenesis. Clonal fertilization, fraternal fertilization, and self-fertilization are possible. Parthenogenesis. Asexual (budding, gemmulation, reduction bodies). Colonial (fragmentation).

A. Class Calcarea

Reproductive cycle—Sponges are dioecious or more often hermaphroditic. They produce gametes which fuse to form a zygote, with nurse cells sometimes also fusing with the ovum. The sperm are sometimes transported by an amoebocyte to the site of fertilization. A flagellated blastula results from cleavage and inversion of the embryo; this amphiblastula swims free of the sponge. Gastrulation occurs and the larva attaches to the substrate as an olynthus, which slowly grows into a sponge.

Basic Bisexual Reproduction occurs as anisogamy.*

Self-fertilization is possible in hermaphroditic sponges, but its occurrence is usually denied.

Clonal fertilization and fraternal fertilization are likely.

Parthenogenesis is possible.

Budding is common.*

Gemmules are not produced.*

Reduction bodies are formed in some species during adverse conditions.*

Colony duplication occurs by fragmentation and regeneration of branches which are broken off.* However, we find no direct statement that the breaking is self-imposed; it usually appears to be accidental.

B. Class Hexactinellida

Reproductive cycle—These hermaphroditic sponges produce gametes, which fuse to form a zygote, which develops into a flagellated larva, although the processes are poorly known. The larva settles and develops into a sponge. Some species also reproduce by buds and gemmules, and self-fertilization cannot be excluded.

Basic Bisexual Reproduction occurs as anisogamy.*

Self-fertilization is possible in hermaphroditic sponges, but its occurrence is usually denied.

Clonal and fraternal fertilization are likely.

Budding occurs in some species, with the buds arising from almost any part of the surface.*

Gemmules appear to be formed, although the process has not been followed.*

C. Class Demospongia

Reproductive cycle—These hermaphroditic sponges produce gametes, which fuse to form a zygote, which develops into a flagellated stereogastrula. This escapes, in some forms everts, and attaches at its anterior end. It then gradually develops into a sponge. Budding, gemmulation, and reduction bodies may interrupt this cycle. In *Tethya maza* budding may succeed budding occasionally.

Basic Bisexual Reproduction occurs as anisogamy.*

Self-fertilization—As most sponges are hermaphroditic, self-fertilization is possible. Usually eggs and sperm are produced at different times.*

Clonal and fraternal fertilization are likely, especially in colonies.

Budding occurs in *Tethya maza,* with secondary buds sometimes forming on the bud before separation.*

Gemmules are produced in many species, especially in the Spongillidae. The latter hatch directly into young sponges, whereas the gemmules of marine sponges give rise to flagellated larvae.*

Reduction bodies have been observed* in both marine and fresh-water species. They arise as indefinite masses of amoebocytes surrounded by epidermis, formed when a sponge disintegrates in unfavorable conditions. They can give rise to many new sponges.* In some species these may not be multiple (and therefore may not be reproductive).

D. Class Sclerospongea

This recently recognized class consists of six genera, some of which were previously known, and may also include the extinct Stromatoporoidea. No description of the reproduction of members of this group has been available to us, and it can only be guessed that they will show sexual reproduction, budding, and possibly gemmulation of some sort.

IV. PHYLUM MESOZOA

Habits—These are parasites of a variety of invertebrates, including flatworms, nemerteans, clams, cephalopods, annelids, and brittle stars, living in ducts, internal spaces, or tissues. The life cycles are not fully worked out, but the structure is essentially that of a blastula, with no digestive tract or other internal cavity. They are unique among metazoans in reproducing by agametes. None form colonies.

Reproduction—Parthenogenesis is possible. Asexual (agamogony, fission, division, fragmentation). (Recent articles fail to complete the cycle and have not clarified the reproduction. The present interpretation is that of Hyman.)

A. Class Dicyemida

Reproductive cycle—There is a break in the known cycle, during which it is presumed that a sexual phase occurs, from which arises the ciliated larva which is the first known stage. This larva develops into a stem nematogen, inside which one agamete forms from one of the axial cells. This agamete then divides into many, each of which then develops (by ordinary cleavage) into another nematogen which will produce more agametes. This cycle continues during the growth of the cephalopod host. Each nematogen then changes slightly into a rhombogen (or the rhombogen may develop directly from an agamete). The axial cell of the rhombogen produces agametes by fission; these cleave into balls of cells or infusorigens. Cells are given off from the surface of the latter; these are called pseudo-eggs. They develop into free-swimming ciliated infusoriform larvae, which leave the mother rhombogen and disappear. It is believed that they initiate a sexual cycle which results in the ciliated larva that infects a new host.

Basic Bisexual Reproduction does not occur, and no sexual individuals have been identified.* It is possible that the supposed "sexual" phase is parthenogenetic.

Parthenogenesis may occur in the unknown part of the cycle (assumed to be sexual). It is definitely reported in *Dicyemennea schulzianum,* where the spermatozoan nucleus is cast off by the activated egg.

Agamogony occurs through fission of each of the three axial cells of the stem nematogen and probably additional fissions of the fragments, to form many agametes. Each agamete cell finally develops inside the stem nematogen to form a daughter (or ordinary) nematogen or sometimes a rhombogen.*

Fission occurs in the unicellular agametes inside both the nematogens and the rhombogens.*

Division (or budding) occurs in the infusorigen stage, when surface cells separate off to become pseudo-eggs (agametes of a distinct sort).* (Because these animals consist of about 31 cells in all, it is reasonable to label the separation of one surface cell as division; it has been called budding, but there is no proliferation of cells for this purpose.)

B. Class Orthonectida

Reproductive cycle—The infective stage is a plasmodium, probably corresponding to the axial cells of the stem nematogen of dicyemids. The plasmodium fragments (agamogony) into agametes that cleave into morulae that are similar to nematogens. These are the sexual phase, being male, female, or hermaphrodite, the axial space of each being filled with ovocytes, spermatogonia, or both. The gametes are fertilized within the female after copulation, and the fertilized eggs develop inside into ciliated embryos resembling infusoriform larvae. They escape into the sea and infect new hosts, when they lose their somatic layer and the interior cells scatter to form many plasmodia.

Basic Bisexual Reproduction does not occur. Although the species are usually dioecious,* and fertilization is internal through copulation, asexual processes always intervene.

Self-fertilization—Hermaphroditic species occur, but self-fertilization has not been reported.*

Agamogony occurs inside the ciliated larva that arises from the zygote when it enters its new host; it loses its somatic cells, and the agametes each give rise to a plasmodium.* It also occurs in the plasmodium.

Fragmentation occurs in a plasmodium to produce agametes, and in the ciliated larva to produce many cells that grow into plasmodia.

V. PHYLUM MONOBLASTOZOA

Habits—The only known species, *Salinella salve,* is an aquatic animal living in saline water but seen only once, in 1892. It consisted of a single layer of cells around a digestive tube with mouth and anus. It was ciliated and presumably could creep. It apparently lived on detritus. Colonies are not known.

Reproduction—(Sexual processes may include hologamy.) Asexual (fragmentation possible, division).

Reproductive cycle—The presumed adult of *Salinella* is a one-layered cylinder ciliated both inside and out. It reproduces by transverse division. Two individuals may encyst together, but the outcome of this possibly sexual process is not surely known. Ciliated unicellular young may have come from the cysts.*

Basic Bisexual Reproduction does not occur, because asexual processes always intervene. Two individuals have been seen to adhere and encyst together, and ciliated unicellular larvae may result from this union.*

Division (transverse) of the adult occurs.*

VI. PHYLUM PLACOZOA

This recently recognized group consists of two genera (possibly only two species) that had previously been placed in the Mesozoa, where they surely do not belong, or in the Coelenterata (Hydrozoa). Their similarity to planula larvae gave rise to the latter assignment, but there are no definitive characters showing them to be coelenterate. It includes *Trichoplax.*

Information on their reproduction is not available to us, and suggestions about sexual reproduction in recent literature are not supported.

VII. PHYLUM COELENTERATA

Habits—They are mostly marine, but one class, Hydrozoa, includes many freshwater species also. The animals may occur in three forms, polyp (usually attached), medusa (always free-swimming), or floating polymorphic colony (in addition to developmental stages). The polyps may also be colonial. A species may consist only of polyps, only of medusae, of both in sequence in the life cycle, of both simultaneously, of a composite colony of definite shape containing individuals of as many as seven forms of polypoids and medusoids, or of a colony of indefinite shape and size consisting of similar polyp individuals. A few are ectoparasites of fish and snails, and some are food stealers as larvae.

Colonies are of several sorts. Polyps may bud other polyps to form a branching colony; strobilation may produce a temporary colony like a stack of saucers; a stem polyp may bud the polymorphic individuals of a siphonophore floating colony.

Reproduction—Basic Bisexual (rarely). Self-fertilization is possible. Clonal fertilization occurs in colonies of polyps. Asexual (division, pedal laceration, polyembryony, budding, frustulation, strobilation, podocysts). Colonial (fragmentation of entire colonies, eudoxy).

A. **Class Hydrozoa**
    Reproductive cycle—A variety of processes occurs in an unending variety of sequences or alternatives. In general the species are dioecious or rarely monoecious. The individuals produce gametes, which fuse to form a zygote, which develops into a planula larva. This usually develops into a polyp. Budding is widespread, producing polyp from polyp; medusa from polyp via a gonophore; medusa from medusa via a blastostyle or directly by budding or division; actinula larva from actinula; planula larva from planula. The buds may occur on stolons, hydrocaulus, hydranth stalks, hydranths, or blastostyles. Division and fragmentation occur in many species and stages. The following are examples of some sequences:

Hydra—The polyp may bud new polyps directly from its stalk. The polyp may develop ovaries or spermaries which produce gametes. Fusion of these produces a zygote, which develops into a planula larva, and the larva attaches and develops into a polyp. This bud apparently arises solely from epidermis. The bud develops endoderm by differentiation, and this later becomes continuous with the endoderm of the parent.*

Margellium—The medusa may produce buds on the manubrium; they arise from epidermis on the interradii; the epidermis laminates to form endoderm around a coelenteron; an invagination of epidermis later produces the subumbrellar cavity of the future medusa. This is called germinal budding, because it arises from so-called germinal epithelium, the site of future gonads in the epidermis;* the buds may themselves develop buds before liberation.

Cunina—In individuals of either sex, undifferentiated germ cells in the epidermis become amoeboid and wander into the endoderm. They divide into two cells, one of which becomes enveloped by the other. The outer cell continues without division to provide protection and nourishment to the inner cell. The latter, called a spore cell by some writers, divides and eventually gives rise to a larva which develops into a medusa. This process has been called sporogony;* it is not the same as sporogony in Protozoa and is better labelled as agamogony, as the germ cell is reproductive in nature but is not a gamete; therefore it is an agamete.

Obelia—The initial polyp buds feeding polyps (hydranths) until a branching colony is formed, then reproductive polyps (gonangia) are budded, from which are budded medusae; the medusae are sexual, producing gametes, which fuse to form a zygote. The planula larva develops into a polyp.

Aeginopsis—A medusa produces gametes, which fuse to form a zygote, which develops into a planula larva. This briefly forms a polyp or hydrula which develops into a medusa.

Liriope—The medusae produce gametes, which fuse into a zygote. The resulting planula larva develops into a tentaculate actinula larva and then into a medusa.

Siphonophora—The gonozooids produce gametes, which fuse to form a zygote, which develops into a planula larva. From the side of the latter buds a swimming bell, and then the larva becomes a feeding polyp or gastrozooid. A stem grows out between and then buds other gastrozooids as well as dactylozooids (food capturing and protective polyps) and gonozooids (reproductive polyps). The pneumatophore, or float, is also budded from the stem. The gonozooids start the cycle over again. There may be as many as three kinds of polypoids and four kinds of medusoids in one colony.

Basic Bisexual Reproduction can occur as anisogamy, but asexual processes are so common as to make this unlikely.

Self-fertilization may occur in *Eleutheria,* in which eggs and sperm develop together in the walls of a brood pouch.* Asexually-produced colonies may be hermaphroditic; in these, syngamy could be apomictic (clonal).

Division, longitudinal, is known to occur in hydrozoan medusae.* Both transverse and longitudinal fission occur in polyps such as *Hydra.** In a few cases, longitudinal division is the only method of reproduction.* In transverse "fission" (as it is usually called) a constriction forms, the parts separate, and each half regenerates.* In longitudinal "fission" the division starts at the hypostome and proceeds to the stem; by the time the two are connected by only a small band of tissue, each daughter polyp possesses all the organs of the original individual.*

Fragmentation occurs in the form of pedal laceration, in which the polyp leaves behind fragments of the foot as it moves.* In *Hypolytus* a stolon may break up into reproductive bodies inside the periderm, with the bodies then breaking off and growing into new polyps.*

Polyembryony probably occurs, as single blastomeres or groups of them can develop normally like whole eggs.*

Budding of new polyps in colonial forms occurs on stems or stolons, or in the single case of *Heterostephanus* on the hydranths.* These buds result in new hydranths (feeding polyps) or gonangia (reproductive polyps) or as many as seven types of polypoids and medusoids (Siphonophora).* In solitary hydroids, the buds may form on the side of the polyp, although this area is usually called the stalk.* Medusae are budded from the blastostyle of gonangia* or from gonophores budded from the sides of polyps.* Medusae may also be budded from medusae, especially in Anthomedusae.* The buds may form on manubrium, bell margin, tentacular bulbs, radial canals, etc.*

Budding may occur in the larval stage, as when the actinula of Trachylina buds other actinulae* or the planula of the Calycophora buds the other members of the colony.* *Moerisia* produces a type of double bud in which the larger bud is transformed into a polyp, whereas the smaller bud becomes a medusa.* In the medusa of *Epenthesis macgradii,* some buds give rise not to medusae but to blastostyles from which medusae later are budded.* In some Narcomedusae, the larvae produce "ramified arms" on which a large number of buds appear, from which whole clusters of medusae develop.*

Frustulation involves the constricting or budding off of planula-like pieces from the side of the polyp, in such forms as *Microhydra* and *Craspedacusta,* to form nonciliated bodies that creep or lie on the bottom and develop into polyps.* It also occurs in *Schizocladium,* where the frustules have been termed "spores" and the process "sporogony."*

Colonies may be duplicated by branches constricting off to develop into new colonies.* Stolons often detach to set up independent colonies.* In some Siphonophora, colonies fragment by a process called eudoxy, separating off cormidia, to produce daughter colonies.*

B. Class Scyphozoa

Reproductive cycle—In the usual dioecious species, the medusae produce gametes, which fuse to form a zygote, which develops into a planula larva. These larvae may bud off other planulae by means of stolons; they eventually transform into polypoid scyphistomae. Transverse septae convert the scyphistoma into a strobila (polydisk strobilation) or the ephyrae are formed one at a time and liberated (monodisk strobilation). The ephyra is a larval medusa, and it develops into a sexually mature medusa.

Basic Bisexual Reproduction probably does not occur, because strobilation is universal (except in *Pelagia*,* where other possible processes are not mentioned). Hermaphroditism is rare.

Self-fertilization is possible in the few hermaphroditic forms such as *Chrysaora*, but it has not been reported.*

Fragmentation occurs in the form of pedal laceration, when the scyphistoma leaves behind small pieces that can regenerate.* Cast-off tentacles, in *Chrysaora*, may round up into ciliated pseudoplanulae, then develop into scyphistomae.*

Budding produces new planulae from stolons from previous planula larvae (in some of the Stauromedusae).* In the Semaeostomeae, a scyphistoma larva may bud off another scyphistoma from a stolon.* Scyphistomae may also bud from the side of the stalk, and the constrictions of the strobila may produce scyphistomae rather than ephyrae.* The scyphistoma stalk may bud off planulae.* In the Lucernariidae, a planula, attached by its anterior end, may bud at the posterior end four planula-like larvae.*

In *Cassiopeia*, the upper part of the scyphistoma stalk constricts off ciliated buds, which resemble planulae and develop like them into new scyphistomae.* A stereogastrula may bud planulae.*

Strobilation occurs in the scyphistoma (larval) stage, either as successive production of medusae (monodisk) or as continuous constriction into ephyrae which do not escape until several are present in a chain (polydisk).*

Podocysts are secreted by the pedal disk, enclosing epidermal and mesenchymal cells. They hatch into ciliated larvae.*

C. Class Anthozoa

Reproductive cycle—The sexual polyps may be dioecious or hermaphroditic; they produce gametes, which fuse to form a zygote, which develops into a ciliated planula larva. This larva eventually settles to the bottom, develops tentacles on one end, and becomes a polyp. This normal sequence is interrupted by various forms of division and budding, the latter being the method of origin of all colonial individuals after the first.

Basic Bisexual Reproduction can occur as anisogamy in solitary species. Asexual processes would generally interfere.

Self-fertilization, although not reported, is possible. Although the individuals are often protogynous or protandrous, the colonies may be hermaphroditic with resulting likelihood of clonal fertilization.*

Clonal fertilization occurs in colonial forms.

Division in the anemones may be either transverse or longitudinal. It is transverse only in the primitive *Gonactinia prolifera*, where a second circlet of tentacles develops on the column of a young polyp, which then constricts in two above the new tentacles.* Longitudinal division occurs in *Sagartia* and some Madreporaria, in which the pedal disk widens until it splits, the rupture proceeding up the stalk to the oral disk. The rupture may be less regular and result in more than two pieces; the torn edges grow together.* It also occurs in the planula of actinians.*

Fragmentation occurs as pedal laceration in such genera as *Metridium* and *Heliactis*. Lobes are constricted off from the pedal disk or pieces of the disk adhere to the substrate and are torn off as the animal moves about. These fragments may undergo further fission before developing into several new individuals.* Cast-off tentacles may regenerate into new anemones.*

Budding occurs from the polyp in *Gonactinia*,* from the pedal disk in *Actinea*,* from a tentacle in *Boloceroides*,* from a stolon in Stolonifera,* or from a solenium, gastrodermal tube between polyps, as in Alcyonacea and Gorgonacea.* In the colonial corals (Madreporaria), budding produces new mouths and then new individuals either outside or inside the ring of tentacles.*

Strobilation (monodisk) occurs in *Fungia* by cutting off the oral ring repeatedly, with regenerations.*

VIII. PHYLUM CTENOPHORA

Habits—All are marine and all are free-swimming except the members of the four genera of the order Platyctenea, which are flattened for a creeping existence. In one of these *(Gastrodes)*, an early larval form is parasitic in tunicates. Colonies are not formed.

Reproduction—Basic Bisexual. Self-fertilization is possible. Asexual (division, fragmentation as pedal laceration). Dissogeny is said to occur in many ctenophores, but the species are not identified.

A. Class Tentaculata

Reproductive cycle—The hermaphroditic adults produce gametes, which fuse into a zygote. The larva is a cydippid, which in Cydippida must change only slightly to become adult.* In Lobata and Cestida there is a substantial metamorphosis to the adult form. In *Gastrodes* alone, the zygote develops into a planula larva, which burrows into a tunicate, where it changes into a cydippid larva.*

Basic Bisexual Reproduction surely occurs, but universal hermaphroditism and occasional brooding make this unsure in many cases; and the occurrence of a second period of sexual activity in the larva is too little understood to be analyzed for amphimixis. In *Ctenoplana*, only testes have so far been found.*

Self-fertilization—Hermaphroditism is universal, with possibility of self-fertilization in the gastrovascular cavity.* Pelagic species are known to be self-fertile.*

Division—It is suspected that ctenophores may reproduce by division.*

Fragmentation (pedal laceration)—In the creeping *Coeloplana* and *Ctenoplana*, small portions are left behind as the animal moves, regenerating later into complete new individuals.*

B. Class Nuda

Reproductive cycle—The hermaphroditic adults produce gametes, which fuse to form a zygote, which develops into a cydippid larva that possesses no tentacles; it changes to an adult largely by a great expansion of the stomodaeum.*

Basic Bisexual Reproduction probably occurs.

Self-fertilization could occur in *Beroe*, which is hermaphroditic.*

Asexual reproduction is not known.

IX. PHYLUM PLATYHELMINTHES

Habits—The flatworms are flattened worms with a branched digestive tract or none at all. They are free-living (most Turbellaria), ectocommensal (some Turbellaria), ectoparasitic on Crustacea (a few Trematoda), or endoparasitic (a few Turbellaria, most Trematoda, all Cestoda, and all Cestodaria). Among the free-living species, the Turbellaria are marine or freshwater, the Gnathostomulida are interstitial marine.

Colonies of a sort are formed in the Turbellaria by transverse division into chains of zooids. The strobila formed by Cestoda are not chains of individuals but merely chains of gonophores and thus not colonies in any sense.

Reproduction—Basic Bisexual. Self-fertilization. Parthenogenesis (pseudogamy). Asexual (division, fragmentation, frustulation, budding in larval stages, polyembryony, successional polyembryony).

A. **Class Gnathostomuloidea**

Reproductive cycle—The hermaphroditic adults produce gametes, which fuse to form a zygote. The sperm are of several strikingly different types, some with and some without flagellar tails. Apomixis is not known for sure.

Basic Bisexual Reproduction presumably occurs in all, although other processes cannot be excluded.*

Self-fertilization—Protandry has been suggested as possibly occurring in some, and copulation apparently occurs.* It is not yet possible to exclude self-fertilization in all.

Parthenogenesis cannot be excluded, although not reported.

Asexual reproduction—Fragmentation of the body is apparently common in some species. Most fragments collected are the anterior end, and it is believed that only this head end can completely regenerate. (The fragmentation may be associated with oviposition.*)

B. **Class Nemertodermatida**

Reproductive cycle—The hermaphroditic individuals produce gametes which fuse to form a zygote. Cross-fertilization appears most likely, because spermatozoa are liberated and ova retained.*

Basic Bisexual Reproduction is likely.*

Self-fertilization is possible but unreported.

Parthenogenesis is possible but unreported.

Apomictic reproduction is not reported, although it is not possible to exclude either self-fertilization or parthenogenesis.

C. **Class Xenoturbellida**

Reproductive cycle—Apparently hermaphroditic, ova and spermatozoa are produced together "all over the parenchyma."* Details of development and larval stages are not available to us.

Basic Bisexual Reproduction may occur, as fertilization is said to be "probably external."*

Self-fertilization cannot be excluded.

Asexual reproduction is not reported.

D. **Class Turbellaria**

Reproductive cycle—Hermaphroditic adult turbellarians generally copulate and produce gametes. These unite to form a zygote, which develops into young worms (juveniles). Especially in the Typhloplanidae, two kinds of eggs are produced, thin-shelled or subitaneous eggs and thick-shelled or dormant eggs. Any individual worm may produce both kinds, subitaneous first, or only the dormant eggs.

In such forms as *Dugesia*, each individual has a choice of three processes: (1) bisexual; (2) self-fertilizing; and (3) asexual.

Basic Bisexual Reproduction occurs commonly, but a few species are exclusively apomictic (asexual).* In some Tricladida, compound or composite eggs are formed around 2 to 40 zygotes; the blastomeres from these separate and recombine into a single new embryo.* This unique developmental process results in individuals that have mixed genomes.

Self-fertilization is rare, even though hermaphroditism is almost universal.* In general, dormant eggs are amphimictic; subitaneous eggs are self-fertilized.* In *Diopisthoporus*, male, female, and hermaphroditic individuals may occur simultaneously.*

Pseudogamy occurs in Tricladida, especially with polyploidy.*

Parthenogenesis occurs and may be obligate.* Gonomery occurs.*

Division, transverse, after duplication of most organs, occurs only in a few forms, such as some *Planaria* and *Stenostomum;* in the rest (as in *Dugesia*) there is little indication of new organs before division.* The worm divides into two parts, each of which may divide again. Transverse division forms chains of individuals in the Catenulida,* but this is not strobilation because any zooid in the chain may divide.

Fragmentation into many small pieces occurs in *Phagocata velata,* with the process either repeating indefinitely or alternating with a sexual phase.*

Frustulation is said to occur.

E. Class Temnocephaloidea

Reproductive cycle—These animals are commensal on freshwater animals; they are hermaphroditic with a common gonopore. The adults produce gametes and a zygote, from which juveniles hatch.

Basic Bisexual Reproduction may occur, but the structure of the reproductive ducts leaves this uncertain.*

Self-fertilization may be possible because there is a common antrum and a single gonopore.*

Asexual reproduction is not reported.

F. Class Trematoda

Reproductive cycle—The cycles of these parasitic organisms vary greatly in details, largely because of different numbers of hosts. In general in the endoparasitic Digenea, the hermaphroditic adults produce gametes, which fuse to form a zygote, which develops into a ciliated miracidium larva. The larva enters the tissues of the host (usually a mollusk), where it either produces rediae directly or changes into a sporocyst, which then produces many rediae or cercariae. The rediae produce cercariae, which may either go directly to the final vertebrate host, or change to a metacercaria (1) on vegetation; (2) in a different host; or (3) while still in the mollusk. Upon reaching the final host, the parasite matures into an adult.

In the ectoparasitic Monogenea, the dioecious adults produce gametes, a zygote, and a ciliated larva (onchomiracidium or, after fusion, diporpa), which attaches to the host. It develops gradually into an adult. *Polystoma integerrimum* is hermaphroditic, but the cycle is similar to that of *Gyrodactylus*. The larva in *Sphyranura* is really a juvenile. A few Monogenea are hermaphroditic and show no asexual reproduction.

The following are examples of the cycles:

Gyrodactylus elegans—Two or more fertilized ova fuse and then develop into one larva, but in this genus the new individual may contain an embryo, inside which is already a third embryo, inside which is a fourth embryo. The origin of these is called polyembryony, but nothing in the nature of simultaneous multiple division of zygote or embryo is described. It seems more likely that they form in sequence from a germ cell line that is maintained distinct in each embryo. If so, it is what here is called successional polyembryony and is surely asexual.

Parorchis acanthis—The fertilized ovum undergoes cleavage, but from the first division a single propagatory cell remains distinct; further cells arise from the so-called "ectoderm cell" formed with the propagatory cell at this first cleavage. As the early embryo develops, the propagatory cell cleaves several times more, each time with one half becoming part of the developing embryo and the other remaining as a propagatory cell. A small cavity then appears in the posterior region of the embryo, and the germ cell moves into it. It now behaves exactly as the original fertilized ovum did, dividing into two cells of which one cleaves to form an embryo and the other remains a propagatory cell. This forms a second embryo inside the first, containing a single propagatory cell.

As the embryo becomes a miracidium larva, it thus already contains a larva of the next stage, redia, in which the germ line is maintained as a mass of germ cells. Now, a second generation of rediae are formed in the same manner from the cells of the germ balls. Inside these daughter rediae, new germ balls of the same line are forming daughter cercariae in the same manner. The cercaria enters a new host to encyst, where it transforms into a metacercaria. When ingested by the final host, it develops into an adult.

This seems to be a sort of successional polyembryony, as the successive larval stages are not derived by metamorphosis or division of the previous stage but by fissions of the original ovum and its germinal daughter cells.

Fasciola hepatica—As in *Parorchis,* except as follows: The miracidium contains a mass of germinal cells; when it enters the molluskan host it becomes transformed into a sporocyst full of the expanding germ balls. Each germinal cell divides and grows to form a redia and its germ cells. When the redia is liberated, its germ cells each divide as before to form a cercaria and its germ cells. The cercaria swims free from the snail and soon settles on a plant, where it forms a cyst, inside of which it becomes a metacercaria. These germ ball divisions are herein called successional polyembryony.

Basic Bisexual Reproduction occurs generally. In *Diplozoon* two diporpa larvae unite into an X-shape and grow permanently together.*

Self-fertilization—Hermaphroditism is the usual condition (except in Didymozoonidae and Schistosomatidae).* Although cross-fertilization is usual, self-fertilization by sperm wandering or autocopulation is possible* and has been reported in both Digenea and Monogenea.*

Polyembryony—The production of thousands of flukes of one sex from one miracidium (in Schistosomatidae) suggests that polyembryony is the process involved.* Germ balls in the sporocysts in the Digenea may multiply by dividing into other germ balls (embryos).* See also *Gyrodactylus,* the first example cited.

G. Class Cestoda

Reproductive cycle—The hermaphroditic adults, in each mature proglottid, produce gametes which fuse to form a zygote which is enclosed in membrane with yolk cells, all covered by a "shell." In some groups this outer covering is ciliated. In the egg there develops an oncosphere or hexacanth larva. Development from here to the adult stage passes through one or more of the following stages: procercoid, plerocercoid, cysticercoid, cysticercus (bladder worm), caenurus, and hydatid. In the final host, the cysticercoid or the bladder worm (or its counterpart in the caenurus or hydatid) everts a scolex, the original and anchoring part of the adult tapeworm. The lower end of this grows and forms transverse septae, with the isolated region in each case growing into a proglottid with complete organ systems. By continuing growth and repeated septum formation at the same point, a chain of proglottids is formed with the most distant proglottid being the oldest. This is strobilation, but the proglottids are multiple gonophores rather than separate individuals. Some asexual multiplicative processes appear irregularly in these cycles. (It must be remembered that a few "tapeworms" do not strobilate and thus do not produce a chain of proglottids.)

Basic Bisexual Reproduction can occur only accidentally, involving two tapeworms and would affect only one proglottid. More often hermaphroditism results in clonal or self-fertilization within one strobilus or one proglottid. Species of only one genus (*Dioecocestus*) are dioecious.*

Self-fertilization—Hermaphroditism occurs in every proglottid. Fertilization between proglottids would not be apomictic but comparable to fraternal fertilization, and self-fertilization within each proglottid probably occurs most commonly.*

Parthenogenesis is likely and apparently has been reported in the genera *Diplophallus*, *Gyrocoelia*, and *Zufala*, but documentation is not available.

Division of a cysticercus into two or three may occur.*

Fragmentation—In Pseudophyllidae, plerocercoids may undergo fragmentation, but apparently only the fragment bearing the hooks can regenerate, so this does not produce new individuals.*

Budding occurs in these stages: coenurus bladder,* hydatid cyst of *Echinococcus*,* and cysticercus.*

In *Sparganum prolifer*, some multiplication takes place in abnormal hosts, apparently at the plerocercoid stage.* In Taenioidea, cysticercoids and cysticerci may bud externally or internally to form more of the same stage.* In *Urocystis*, an ingested oncosphere may bud cysticercoids. Repeated budding of the cysticercus occurs in *Taenia crassiceps*.* In coenurus and hydatid bladders there may be enormous multiplication of scolices through both internal and external budding.* In *Urocystidium* there is proliferation of young strobila as outgrowths from the end of a parent strobilus.*

Strobilation—Although formation of a strobilus (chain of proglottids) occurs, it does not in itself produce new individuals,* but merely produces multiple gonophores.

Polyembryony is said to occur in *Echinococcus* and *Multiceps*.*

H. Class Cestodaria

Reproductive cycle—The cycles of most are incompletely known. The adults are hermaphroditic. They produce gametes, which fuse to form a zygote, which develops into a larva with ten hooks and sometimes cilia, called a lycophore or decacanth. The larva may pass through procercoid and plerocercoid stages before reaching maturity in the final host. Strobilation does not occur.

Basic Bisexual Reproduction presumably occurs.

Self-fertilization is at least possible because hermaphroditism is universal.*

Asexual reproduction is not reported.

X. PHYLUM RHYNCHOCOELA

Habits—These elongated worms are mostly free-living and marine, living on the littoral ocean bottom or being bathypelagic. A few are commensal. Some occur in freshwater and terrestrial habitats. None are colonial.

Reproduction—Basic Bisexual. Self-fertilization is possible. Parthenogenesis is possible. Asexual (fragmentation).

Reproductive cycle—The species are dioecious or hermaphroditic, and the adults produce gametes. These fuse to form a zygote which develops into a larva, which is either a juvenile (maturing directly into an adult) or a Desor's larva or a pilidium larva. The Desor's larva is a ciliated post-gastrula, which metamorphoses into an adult. The pilidium is a free-swimming ciliated larva, which also metamorphoses. Asexual processes occur only in some species of *Lineus*.

Basic Bisexual Reproduction predominates in both dioecious and monoecious species. At least *Lineus sanguineus* is not known to reproduce sexually.

Self-fertilization is possible in some of the hermaphroditic species but may not occur in nature. There may be a combined ovotestis.*

Parthenogenesis is possible although unreported.

Fragmentation occurs in many forms, but regeneration of fragments other than the head occurs only in some species of *Lineus*,* which are thus the only cases in which it is multiplicative.

XI. PHYLUM ACANTHOCEPHALA

Habits—These elongate worms are all endoparasites in the digestive tracts of fishes, reptiles, birds, and mammals. They are marine, fresh-water, or terrestrial. The larva is parasitic in an arthropod such as an insect, amphipod, or isopod. None forms colonies.

Reproduction—Basic Bisexual occurs. Parthenogenesis is possible.

Reproductive cycle—The species are all dioecious, and the adults copulate. They produce gametes, which fuse to form a zygote which develops into an acanthor larva in a heavy shell. Upon entering the intermediate host, the larva metamorphoses into an acanthella, which later becomes a juvenile worm. Because of eutely (cell constancy) no ordinary asexual processes are possible in the adult.

Basic Bisexual Reproduction is apparently universal, with the sexes always separate,* but parthenogenesis is also possible.

Apomixis has not been reported in any form and fission and budding are impossible in adults because of eutely. Parthenogenesis cannot be excluded.

XII. PHYLUM ROTIFERA

Habits—The microscopic wheel-animalcules are sessile or free-swimming. They may be epizoic or even ectoparasitic, on Crustacea and Oligochaeta, or endoparasitic in aquatic fly larvae, heliozoan Protozoa, snails, algae, *Volvox* colonies, hydrozoans, Oligochaeta, and slugs. A few are marine or are found in brackish water, but most live in fresh water. Some are classed as terrestrial because they live in moss and lichens and even on glaciers, roofs, rocks, and trees, but in all such cases they live in water films. None form colonies.

Reproduction—Basic Bisexual and parthenogenesis occur.

A. Class Seisonidea

Reproductive cycle—The dioecious adults copulate. They produce gametes, which fuse to form a zygote, and the egg hatches into a young adult. No asexual processes occur, but parthenogenesis is possible.

Basic Bisexual Reproduction seems to be universal; all species are dioecious.* Only one type of egg is produced.*

Apomixis—None has been reported and no asexual reproduction is possible in adults because of eutely.

B. Class Bdelloidea

Reproductive cycle—All individuals are females. They lay parthenogenetic eggs that hatch into young adults. No sexual processes occur.

Basic Bisexual Reproduction does not occur, as males are unknown.*

Parthenogenesis—No males are known for any species,* so parthenogenesis is universal.

Asexual reproduction in adults is impossible because of eutely. Some regeneration is possible in young individuals, but it presumably never produces new individuals.*

C. Class Monogononta

Reproductive cycle—Although all species are dioecious, males may be absent or only seasonally present. Three types of eggs are produced: thin-walled amictic eggs which develop parthenogenetically into females, smaller thin-walled mictic eggs that develop into males if not fertilized, and thick-shelled dormant eggs that are simply fertilized mictic eggs and invariably hatch into mictic females. Any given female is either mictic or amictic and does not produce both kinds of eggs. At hatching, the rotifer is a juvenile or young adult.

Basic Bisexual Reproduction occurs, but it is not universal.* As the species are all dioecious and copulation occurs, presumably all fertilization is amphimictic, but it is not universal.*

Partheonogenesis occurs in the formation of amictic eggs which develop into females.* So-called mictic eggs if not fertilized develop into males; if fertilized, into amictic females.*

Asexual reproduction in adults is impossible because of eutely. Some regeneration is possible in young individuals, but it presumably never produces new individuals.*

XIII. PHYLUM GASTROTRICHA

Habits—The Macrodasyoidea are all marine; the Chaetonotoidea are either marine or fresh-water (commonly). All are free-swimming. Colonies are not formed.

Reproduction—It is Basic Bisexual. Self-fertilization is possible. Parthenogenesis is possible.

A. Class Macrodasyoidea

Reproductive cycle—The species may be dioecious or hermaphroditic. The adults produce gametes which fuse to form a zygote. A juvenile gastrotrich apparently hatches from the egg.

Basic Bisexual Reproduction probably occurs, and parthenogenesis is possible.

Self-fertilization—Hermaphroditism is almost universal. Some are protandrous, but in some, the male and female pores may be close together or united, making it difficult to deny the possibility of self-fertilization. In *Dactylopodalia,* according to Hyman,* the individuals can be male, female, or hermaphrodite,* but it is not entirely clear whether this is dioecism or some form of successional hermaphroditism.

Asexual processes are lacking because of eutely.

B. Class Chaetonotoidea

Reproductive cycle—As males are entirely unknown, the females produce parthenogenetic eggs, which are of two kinds, subitaneous and dormant. (Remnants of male gonads are present in some species, and in *Neodasys* and *Xenotrichula* incomplete descriptions leave the possibility of functional hermaphroditism.) The thin-walled subitaneous eggs are produced by young animals under favorable conditions, the thick-walled dormant or winter eggs by older females in less favorable environments. Both hatch into juveniles of adult form.

Basic Bisexual Reproduction is unknown except possibly in *Xenotrichula* and *Neodasys.** In all other species, only females are known, although traces of testes occur in some.

Parthenogenesis is the only method employed (except for possible exceptions noted above), because eutely makes asexual reproduction impossible in adults.*

Asexual processes are lacking because of eutely.

XIV. PHYLUM KINORHYNCHA

Habits—These microscopic marine animals live in bottom debris and get about by worm-like movements. None are parasitic, and none form colonies.

Reproduction—It is Basic Bisexual.

Reproductive cycle—The species are dioecious; the adults produce gametes, which fuse to form a zygote, and the eggs hatch into a larva unlike the adult. A metamorphosis occurs with molting, before reaching adult form.

Basic Bisexual Reproduction seems to be universal. All species are dioecious, and the eggs are presumably always fertilized.*

Asexual reproduction is not possible in adults because of eutely.

XV. PHYLUM PRIAPULOIDEA

Habits—These are rather large marine worms that live in bottom debris. None are parasitic, and none form colonies.

Reproduction—Basic Bisexual occurs.

Reproductive cycle—The species are dioecious; the adults produce gametes which fuse to form a zygote. The egg hatches into a larva which is scarcely more than a juvenile, molting to attain the adult form.

Basic Bisexual Reproduction is assumed, as the species are dioecious,* and gametes are shed into the surrounding water.*

Apomictic reproduction has not been reported, and asexual reproduction is precluded in the adult by eutely.

XVI. PHYLUM NEMATODA

Habits—These are slender vermiform animals, mostly microscopic in size but attaining nearly a foot in length in one species. They are free-living (fresh-water, marine, or terrestrial), plant-parasitic, or parasitic in animals (of virtually all kinds). They never form colonies.

Reproduction—Basic Bisexual occurs. Self-fertilization occurs. Parthenogenesis occurs.

Reproductive cycle—The adults may be dioecious or hermaphroditic and they copulate. They produce gametes, which fuse to form a zygote. The eggs hatch as juvenile worms, which develop with molting to adult size. Even in parasitic species there are no distinct larval forms. In *Trichinella spiralis,* shortening of the life cycle has given rise to suggestion that juveniles are reproductive or that the species is neotenic.

Basic Bisexual Reproduction occurs. As a rule the species are dioecious, with internal fertilization after copulation.*

Self-fertilization—Hermaphroditism (usually protandric) is not uncommon, and self-fertilization has been reported.* Hermaphroditism is sometimes called syngony by nematode workers.

Parthenogenesis has been shown in species of *Mermis, Heterodera,* and *Rhabditis.** In some *Rhabditis* which are dioecious, there is no fertilization of eggs that will become females, merely activation by the presence of sperm.* Pseudogamy (sperm activation without karyogamy) occurs in *Mesorhabditis belari.**

Asexual reproduction is not known and is precluded among adults by eutely.

XVII. PHYLUM GORDIOIDEA (Nematomorpha)

Habits — The adults are always free-living, but the juveniles are always parasitic in arthropods. They are marine (pelagic), fresh water, or terrestrial. None are colonial.

Reproduction — Basic Bisexual.

A. Class Gordioidea

Reproductive cycle — The species are dioecious; the adults produce gametes which fuse to form a zygote. From the egg a larva hatches which enters an arthropod host. It has been called an echinoderid larva, but this fancied resemblance has been rejected by other workers. These become juvenile worms which molt to become adult.

Basic Bisexual Reproduction seems to be universal — always dioecious, with copulation.*

Apomictic reproduction is not reported. Asexual reproduction is precluded in adults because of eutely.

B. Class Nectonematoidea

Reproductive cycle — The species are dioecious; the adults produce gametes which fuse to form a zygote. It is not known whether a larva is produced or merely a juvenile.

Basic Bisexual Reproduction appears to be the rule, as the sexes are separate.*

Apomictic reproduction is not reported.

XVIII. PHYLUM CALYSSOZOA (Entoprocta, Endoprocta, Kamptozoa)

Habits — These are solitary or colonial stalked sessile animals of small but not microscopic size. All are marine, except for one fresh-water genus. Some attach to other animals, but none are parasitic. The colonies are dendritic.

Reproduction — Self-fertilization is possible. Colonies are clonal. Asexual (budding).

Reproductive cycle — The species may be dioecious or hermaphroditic; the adults produce gametes which fuse to form a zygote, with the embryo being brooded in a special chamber. A free-swimming ciliated larva results, which has been called a trochophore but is substantially different. The larva settles and metamorphoses into an adult zooid, then buds other individuals to form a colony.

Basic Bisexual Reproduction does not occur, because asexual processes intervene.

Self-fertilization — Hermaphroditism occurs in a few species, some of these being protandric.* As the gonoducts open into a brood chamber, self-fertilization is possible. Some colonies are unisexual, and all are clonal.

Budding occurs in all forms, from stolons, stalks, or calyces.* In the Loxosomatidae it may start in the embryo still attached in the maternal brood chamber.* Budding produces the colonies in all colonial forms, but in *Loxosoma* the buds are completely detached to start new individuals.*

XIX. PHYLUM BRYOZOA (Ectoprocta, Polyzoa)

Habits — These so-called moss animals are microscopic, sessile, and colonial. The colonies are usually branched but so dense as to appear as an encrusting mat. They are polymorphic and mostly marine. The members of one class (Phylactolaemata) are found only in fresh water, attached to objects. A few bore holes in mollusk shells or live between layers of the tubes of annelids; one lives in the cystids of another bryozoan, which it destroys, and another lives in the tissues of an ascidian.

Reproduction — Self-fertilization, clonal fertilization. Parthenogenesis. Asexual (division, polyembryony, budding, statoblasts, hibernacula).

A. Class Phylactolaemata

Reproductive cycle — The hermaphroditic adults produce gametes which fuse to form a zygote, which they brood in an embryo sac. A ciliated larva results, which is actually a young colony. The embryo has budded other polypides (individuals) from one end.

Basic Bisexual Reproduction does not occur, because fertilization is usually by spermatozoa of the same animal or of the same clonal colony* and budding is universal.

Self-fertilization — Hermaphroditism is universal, and self-fertilization or clonal fertilization seems to be the rule.*

Budding is the method of colony formation, starting in an early embryonic state, so that the ciliated "larva" is actually a juvenile colony of two to four embryonic individuals.* This process has been called polyembryony.* Buds may form on undeveloped buds and even tertiary buds on these — compound budding.*

Statoblasts are minute discoidal resting stages, produced in numbers at certain times. They are internal buds with a chitinous covering that can start new colonies.*

Hibernacula are external buds with thick "sclerotic" coats. They are specially modified winter buds.*

B. Class Gymnolaemata

Reproductive cycle — The species may be dioecious or hermaphroditic; the adults produce gametes which fuse to form a zygote, with the eggs usually brooded in a special chamber or in the coelom. The manner of entry of sperm from other zooids or colonies is not clear, and in some the sperm cannot live more than a few minutes in sea water.* In cyclostomes the embryo cleaves to a morula, puts out lobulations which separate off as separate embryos (polyembryony by budding), some of which may constrict to produce tertiary embryos. In others, from the "egg" hatches a cyphonautes larva, which looks much like a rotifer, which eventually metamorphoses into an adult zooid. This primary zooid, called an ancestrula, buds the further members of the colony.

Basic Bisexual Reproduction does not occur, because even in the dioecious species there is budding.

Self-fertilization — Hermaphroditism occurs in the majority of species. It is sometimes protandrous or protogynous, although apparent protandry may be ineffective, because even a very immature ovocyte may already contain sperm.* In *Bugula,* the eggs leaving the ovary are at once fertilized by spermatozoa of the same zooid.*

Parthenogenesis has been reported in *Crisia* but seems not to have been confirmed.*

Division is known in the larva of *Membranipora,* which divides into two polypides.*

Polyembryony is reported, but it appears to be merely embryonic budding.*

Budding occurs from the ancestrula (zooid),* from stolons,* or from embryos.* In Cyclostomata, the embryo, while still a sphere of several dozen cells constricts off portions which then develop into (100 or more) independent embryos.*

XX. PHYLUM PHORONIDA

Habits — These are tube-dwelling vermiform animals, found only in coastal waters of the sea. They may live in dense masses of tubes but are solitary and never parasitic.

Reproduction — Basic Bisexual. Self-fertilization occurs. Asexual (division), possibly budding.

Reproductive cycle — The adults, usually hermaphrodites, produce gametes which fuse to form a zygote. The latter develops into a ciliated free-swimming actinotroch larva, which passes through a considerable metamorphosis to become adult.

Basic Bisexual Reproduction probably occurs. It has been suggested that the supposedly dioecious species are really protandrous hermaphrodites, but this is also denied.* Fertilization is sometimes in the coelom,* and thus self-fertilization is possible.

Self-fertilization — Hermaphroditism is nearly universal,* but self-fertilization is not reported. However, in species with internal fertilization (in the coelom), such as *Phoronis muelleri,* self-fertilization is likely.

Division, transverse, occurs frequently in *Phoronis ovalis,* with some duplication of organs before separation.* In other cases, regeneration of autotomized parts is common, but the fragments do not become new individuals.*

Budding has been reported but is probably the same process as division, where the new oral end forms at an angle to the original worm.

XXI. PHYLUM BRACHIOPODA

Habits — The lamp-shell animals are all marine and live from the beach to great depth. Approximately 100 times as many species are known as fossils as living. They are solitary but sometimes attached by their shells in masses. Only the larva swims. None are parasitic.

Reproduction — It is Basic Bisexual. Self-fertilization is rarely possible. Parthenogenesis is possible.

A. Class Inarticulata

Reproductive cycle — The species are always dioecious, the adults produce gametes which fuse to form a zygote. The eggs are discharged and fertilized in the sea. When the embryo develops tentacles, it emerges as a free-swimming larva, with cilia only on the tentacles. In *Lingula,* at least, the larva develops into an adult without abrupt metamorphosis.*

Basic Bisexual Reproduction appears to be universal. All species are dioecious.*

Apomictic reproduction is unknown, but parthenogenesis cannot be excluded.

B. Class Articulata

Reproductive cycle — The species are usually dioecious; the adults produce gametes which fuse to form a zygote. The eggs are discharged and fertilized in the sea or are brooded and fertilized internally. When the embryo becomes ciliated, it emerges as a free-swimming larva.

Basic Bisexual Reproduction appears to be universal. Most species are dioecious and sperm are shed into the sea, so self-fertilization is unlikely,* although it is possible in some brooding hermaphroditic species.

Self-fertilization — Hermaphroditism is known only in *Argyrotheca,* *Pumilus,* and *Platidia,* but self-fertilization is not reported.

Parthenogenesis is not recorded but cannot be excluded.

Asexual reproduction is unknown.

XXII. PHYLUM MOLLUSCA

Habits — The mollusks are marine, fresh-water, or terrestrial. They are not polymorphic and never form colonies. A few species of bivalves are parasitic to the extent of living in the intestine of sea cucumbers, and some gastropods appear to be more ectoparasites than predators.

Reproduction — Basic Bisexual. Self-fertilization, Parthenogenesis. Asexual is not known, although polyembryony is said to be sporadic in the phylum.

A. Class Monoplacophora

Reproductive cycle — The species are dioecious, and the adults presumably produce gametes which fuse to form a zygote, but later development is unknown. If these animals are like other mollusks, they produce a trochophore larva.

Basic Bisexual Reproduction presumably occurs, as all are dioecious.*

Parthenogenesis is not recorded but cannot be excluded.
Apomictic reproduction is not known.

B. **Class Amphineura**
Reproductive cycle — With the exception of one species, which is hermaphroditic, these animals are dioecious. The adults produce gametes that fuse into a zygote; each zygote develops into and hatches as a trochophore larva, which metamorphoses into an adult.
Basic Bisexual Reproduction occurs, and all but one are dioecious.*
Parthenogenesis has not been reported but cannot be excluded.
Self-fertilization — Hermaphroditism obtains in *Trachydermon raymondi*, where spermatozoa develop among the ova; therefore it is probably self-fertilizing.*
Asexual reproduction is not known.

C. **Class Solenogastres**
Reproductive cycle — All are hermaphroditic except the Chaetodermatidae, but it is believed that all copulate. Gametes fuse to form a zygote which develops into a ciliated larva (except in brooding species). The larva is sometimes called a trochophore but is different in appearance. Presumably the "trochophore" metamorphoses into an adult.
Basic Bisexual Reproduction presumably occurs, as some are dioecious (e.g., *Chaetoderma*).
Self-fertilization — Hermaphroditism occurs, but it is believed that self-fertilization does not.
Parthenogenesis has not been reported but cannot be excluded.
Asexual reproduction is not known.

D. **Class Gastropoda**
Reproductive cycle — The species are dioecious or hermaphroditic. The adults produce gametes that fuse to form a zygote, which in some marine forms, develops into a trochophore larva. In marine snails the eggs are encapsulated and hatch into an advanced larva called a veliger. In freshwater and terrestrial snails the eggs hatch into juvenile snails, with no larval stage.
Basic Bisexual Reproduction surely occurs, but in marine forms, fertilization in the sea prevents assurance of cross-fertilization in hermaphroditic species.
Self-fertilization — Hermaphroditism is common and gonads may produce egg and spermatozoa simultaneously; yet it is reported that there is no self-fertilization in most groups.* *Lymnaea* seems to be an exception.* Hermaphrodites may be protandrous (Prosobranchia) or simultaneous (Opisthobranchia and Pulmonata), and in the latter may be self-fertilizing.* In *Crepidula fornicata* life begins as a male, passes through an hermaphroditic stage, and ends as a female.*
Parthenogenesis is assumed in some Prosobranchia, because no males have ever been found.* Gonomery occurs.*
Asexual reproduction is not known.

E. **Class Bivalvia**
Reproductive cycle — The species may be dioecious or hermaphroditic; the adults produce gametes, that fuse to form a zygote, which develops into a larva, either a trochophore followed by a veliger or else a glochidium. These metamorphose into adults.
Basic Bisexual Reproduction presumably occurs, as most species are dioecious.*
Self-fertilization — Hermaphroditism occurs in scallops (*Pecten*) where there is a

combined gonad,* as well as in *Ostrea* and *Anodonta*.* Self-fertilization may occur in these. It is definitely known in the Sphaeriidae.*

Parthenogenesis is not reported but cannot be excluded.

Asexual reproduction is not known.

F. Class Scaphopoda

Reproductive cycle — The species are dioecious, and the adults produce gametes, which fuse to form a zygote, which develops into a trochophore larva, with metamorphosis into an adult.

Basic Bisexual Reproduction occurs exclusively, so far as known, with all species dioecious and fertilization external.*

Apomictic reproduction is not known, although parthenogenesis is possible.

G. Class Cephalopoda

Reproductive cycle — The adults may be dioecious or hermaphroditic; they produce gametes which fuse to form a zygote, after copulation or insemination by spermatophore. The embryo develops through a trochophore stage in the egg case, so that a miniature adult is hatched.

Basic Bisexual Reproduction is assumed because of nearly universal spermatophore insemination, but a few species are said to be monoecious.

Self-fertilization — Hermaphroditism occurs in a few species,* but self-fertilization is not reported.

Parthenogenesis is not reported but cannot be excluded.

Asexual reproduction is not known.

XXIII. PHYLUM SIPUNCULOIDEA

Habits — The peanut worms are all marine, living on the bottom in burrows or cavities. None are parasitic, and colonies are not formed.

Reproduction — Basic Bisexual. Self-fertilization possible. Parthenogenesis is possible.

Reproductive cycle — All species are dioecious. Adults produce gametes which fuse to form a zygote, fertilized in the sea. The embryo develops rapidly into a trochophore, that metamorphoses into a worm, which gradually assumes adult appearance.

Basic Bisexual Reproduction occurs. All are dioecious except that hermaphroditism may occur in some species.*

Self-fertilization — Hermaphroditism may possibly occur in a few species.* The spermatozoa, in most cases at least, become active only after ejection into sea water, so direct self-fertilization is accidental at best.*

Parthenogenesis may occur, as males have never been seen in some species.*

Asexual reproduction apparently never occurs, even though regenerative powers are considerable.

XXIV. PHYLUM ECHIUROIDEA

Habits — These are marine worms of moderate size, living in burrows in the bottom or in crevices, in shallow water or in the deep sea. They are solitary detritus feeders, and none are known to be parasitic or colonial.

Reproduction — It is Basic Bisexual.

A. Class Echiurida

Reproductive cycle — The species are all dioecious, and the adults produce gametes, which fuse to form a zygote. The trochophore larva metamorphoses into an adult. Fertilization is external except in *Bonellia,* in which the tiny male lives parasitically in the uterus or coelom of the female.*

Basic Bisexual Reproduction is assumed, as the species are dioecious.*

Apomictic reproduction is not known.

B. Class Saccosomatida

Reproductive cycle — This class, represented by only one known species, is one of the least known of all animal classes so far as general works are concerned. Males are unknown, and apparently nothing is known of the development of these animals.

XXV. PHYLUM MYZOSTOMIDA

Habits — These are small flattened discoidal animals that are known only in the sea in association with echinoderms. Most are commensals or foodstealers, but some are parasitic in the tissues and internal spaces. They do not form colonies.

Reproduction — Basic Bisexual.

Reproductive cycle — The hermaphroditic adults mate and attach a spermatophore each to the other. Thus there are gametes that fuse to form zygotes, and the embryo develops into a trochophore larva. The trochophore metamorphoses into an adult.

Basic Bisexual Reproduction is assumed, as the hermaphroditic individuals inseminate each other by means of a complex, motile spermatophore, as in *Myzostomum,* which includes a motile syncytial carrier inside the receiving individual.*

Apomictic reproduction is not known.

XXVI. PHYLUM ANNELIDA

Habits — The segmented worms, small to very large in size, are marine, fresh-water, or soil-inhabiting. They may be free-swimming, tube-dwelling, or parasitic. A few form chains and branches in temporary or permanent colonies.

Reproduction — Basic Bisexual. Self-fertilization, fraternal fertilization. Parthenogenesis. Asexual (division, fragmentation, strobilation, budding, polyembryony). Colonial (fragmentation of colonies possible).

A. Class Polychaeta

Reproductive cycle — The species may be dioecious or hermaphroditic; the adults produce gametes, which fuse to form a zygote. The trochophore larva metamorphoses into an adult. Several forms of asexual reproduction interrupt this normal sequence, in dioecious species such as those of *Phyllochaetopterus** and *Syllis,* and in monoecious species, such as those of the Sabellidae.* In many of these it is called epitoky (see Figure 13 and definition in the Glossary). The following are examples:

Exogone gemmipara — (not illustrated) The sexually-immature worm produces a posterior chain of partial individuals by a combination of growth and septa formation. The reproductive organs develop in each section, but reach sexual maturity only after the sections become detached. They are always structurally different from the "parent," which also never develops gonads.

Platynereis massiliensis — (Figure 13-1) It may be taken as the simplest type, which does not metamorphose at all; it is hermaphroditic and lays its eggs in its tube, where they

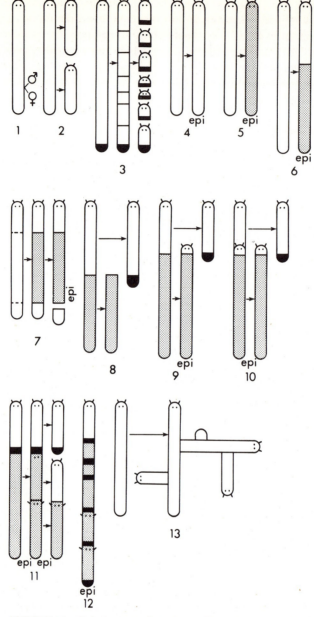

FIGURE 13. Epitoky in the Polychaeta. The original worm is in white; the epitoke in stipple; and regenerating regions in black. See text for individual explanations. (From many sources but inspired by Figure 18 in Dales, R. P., *Annelids*, Hutchinson University Library, London, 1963.)

are brooded. (A closely related species, *P. dumerilii*, does metamorphose, as described below.)

Zeppelina monostyla — (Figure 13-2) It also lacks metamorphosis; it undergoes no sexual reproduction but divides asexually by what seems to be architomy (regeneration after the break).

Sabella spallanzanii — (Figure 13-3) It undergoes multiple fission, first into a chain and then into fragments of various sizes which regenerate the missing parts.

Nephtys sp. — (Figure 13-4) It performs the simplest possible metamorphosis, as only the setae are affected, changing slightly to provide better capability for swimming to the surface for spawning.

Platynereis dumerilii — (Figure 13-5) It metamorphoses throughout its length into a clearly different form for spawning, but there is no division or fragmentation. It dies after spawning.

Nereis irrorata — (Figure 13-6) It has a metamorphosis that affects only the posterior half of the body. Otherwise it is similar to the preceding species.

Tylorrhynchus heterochaetus — (Figure 13-7) It transforms only a center section, leaving the head and tail changed. The latter then breaks off and disintegrates, leaving a partly metamorphosed worm like the preceding species.

Palolo siciliensis and *Syllis hyaline* — (Figure 13-8) They transform by division (architomy), with no regeneration of missing organs. The fragment remains little more than a gonophore, which dies after releasing its gametes. The head end regenerates the tail and survives.

Syllis gracilis — (Figure 13-9) It produces the detached posterior region, but it regenerates the head end and appears as a complete worm, but the division was architomous.

Autolytus sp. — (Figure 13-10) It divides by paratomy, after forming a new head on the rear section.

Syllis vittata — (Figure 13-11) It forms a temporary chain by paratomous divisions, with both the heads and tails formed before separation.

Autolytus prolifer — (Figure 13-12) It forms a chain by strobilation, with all new sections forming from the head piece. The new tails form first and then the heads, leaving the original tail at the posterior end.

Syllis ramosus — (Figure 13-13) It is perhaps outside of this series; it buds lateral individuals that may remain attached and bud others, to form a branching structure.

Basic Bisexual Reproduction occurs, with most species being dioecious.* At least one species of *Zeppelina* has no sexual reproduction.*

Self-fertilization — Hermaphroditism occurs, especially among the fanworms, and in brooding species leave the possibility of self-fertilization in some cases.*

Parthenogenesis has not been reported but cannot be excluded.

Division ("fission"), transverse, is not rare. It occurs as paratomy or as architomy, and is sometimes seen in epitoky (Figure 13), when the gamete-bearing part of the body is separated either as a short-lived fragment or as a separate "individual".* In some Syllidae, the epitoke regenerates a new head and may be considered a new individual – the sexual generation.* There may be more than one epitoke produced, and these may be dioecious or monoecious.* One of the processes of epitoky has been termed schizogamy,* and another schizometamery.*

Fragmentation, sometimes not distinct from division, also occurs, sometimes multiple as in *Phyllochaetopterus*.* At least one species, *Zeppelina monostyla,* lives indefinitely with only this asexual reproduction.*

Budding is usually reported as common, but some cases are more properly division, as defined here. Buds may form laterally, as in *Syllis,** or clusters from one segment as in *Trypanosillis*.*

Strobilation apparently occurs in the Syllidae. This is very diverse and sometimes may not be different from epitokous division.*

Colonial multiplication by fragmentation is possible in chains such as those of *Autolytus* and *Syllis*.*

B. Class Oligochaeta

Reproductive cycle — The hermaphroditic adults inseminate each other and produce gametes which fuse to form zygotes. (It appears that in some cases insemination does not result in syngamy because the spermatozoa cannot reach the ova, so parthenogenesis produces the embryos.) The embryo develops within a cocoon and emerges as a juvenile worm.

Basic Bisexual Reproduction probably occurs, although all species are hermaphroditic.* Some Naididae reproduce only asexually,* as does *Lumbriculus variegatus*.*

Self-fertilization — Hermaphroditism is universal, with the gonads separate except in the Aeolosomatidae.* Self-fertilization is known to occur in a few species, such as *Limnodrilus udekemianus*.*

Parthenogenesis is said to be widespread, even in species that are apparently mutually cross-fertilizing.

Division occurs by paratomy, as in *Nais* and *Aulophorus*,* or successive divisions may take place, forming a chain of individuals.*

Fragmentation, or architomy, either binary or multiple, occurs rarely, as in *Allonais paraguayensis,* where six to eight fragments are formed.*

Polyembryony is reported, but the process described is embryo budding.*

Budding occurs in the embryo of several species, where it is sometimes called polyembryony.* It is also described as division, as in *Lumbricus trapezoides,* where the gastrula divides,* and in aquatic forms such as *Allolobophora caliginosa,* where division is reported both at the first cleavage and at the gastrula stage.* Incompletely separated monsters are common, which distinguishes this process from true polyembryony.

C. Class Hirudinea

Reproductive cycle — The hermaphroditic adults copulate and produce gametes which fuse to form a zygote. Spermatophores are sometimes used in transferring spermatozoa, and there may be a special surface area to receive them. Eggs are laid in a cocoon, and the hatched juveniles may be attached to a parent for weeks or months before they mature into adults. In one genus, *Herpobdella,* there is a definite metamorphosis from a ciliated embryo hatched at an early age.

Basic Bisexual Reproduction presumably occurs, although hermaphroditism is universal.*

Self-fertilization — Hermaphroditism is universal, but the gonads are separate and protandry occurs, and copulation is apparently always practiced; thus, self-fertilization is unlikely.*

Parthenogenesis is not reported but cannot be excluded.

Asexual reproduction is unknown.*

D. Class Archiannelida

Reproductive cycle — Species may be either dioecious or hermaphroditic. Adults produce gametes, which fuse to form zygotes. The trochophore larva metamorphoses into an adult. (There is some chance that cycles of the different genera are more different than now recorded.)

Basic Bisexual Reproduction no doubt occurs in this composite group, although hermaphroditism occurs in some and most accounts give no details.

Self-fertilization — Hermaphroditism occurs in *Protodrilus,** but self-fertilization is not reported.

Parthenogenesis is not reported but cannot be excluded.

Asexual reproduction is not reported.

XXVII. PHYLUM DINOPHILOIDEA

Habits — These very small animals are marine and free-swimming by means of cilia. The males of some are dwarfs. Because they are usually lumped with either Archiannelida or Polychaeta, specific information about them is sparse. Their features connect them with Aschelminthes as much as with Annelida. They are never colonial nor parasitic.

Reproduction — It is Basic Bisexual so far as known to us, except that the genetic diversity that usually results from Basic Bisexual Reproduction is not seen in those species in which siblings cross-fertilize in the cocoon.* The latter is comparable to inbreeding and is here termed fraternal fertilization.

Reproductive cycle — The species are dioecious, and the adults produce gametes which fuse to form a zygote. Insemination is by hypodermic injection. There is no separate larva, and there appears to be no molting.

Basic Bisexual Reproduction presumably occurs, although parthenogenesis cannot be excluded and fraternal fertilization is known.

Fraternal fertilization occurs where the gametes from a pair are deposited in a cocoon, where cross-fertilization by siblings occurs. This process thus approaches clonal fertilization, but the genetic diversity is intermediate between that of a clone and Basic Bisexual Reproduction.

Asexual reproduction is unknown.

XXVIII. PHYLUM TARDIGRADA

Habits — These microscopic animals may be marine (in either shallow or deep water), fresh-water, or terrestrial (but still in water films). They are never colonial nor parasitic. They undergo anabiosis in dry periods.

Reproduction — Basic Bisexual. Parthenogenesis occurs.

A. Class Heterotardigrada

Reproductive cycle — The species are dioecious, and the adults produce gametes which fuse to form a zygote or remain as parthenogenetic eggs. In some species two kinds of eggs are produced, thin-shelled eggs in favorable environments and thick-shelled eggs in adverse environments. Development is direct, the egg hatching into a young adult. Molting occurs four to six times.

Basic Bisexual Reproduction presumably occurs, as all species are dioecious and fertilization is usually internal, although without copulation.*

Parthenogenesis is common.* In genera such as *Echiniscus,* males are unknown.*

Asexual reproduction does not occur, as the anabiosis common in semi-aquatic species is principally dehydration followed by restoration of the original cells and organs,* and eutely is evident in adults.*

B. Class Eutardigrada

Reproductive cycle — The species are all dioecious. The adults produce gametes that fuse to form a zygote, which hatches as a young adult. The eggs of one female may be fertilized by several males, at the time of laying in the shed skin.

Basic Bisexual Reproduction appears to be universal, because the species are dioecious and fertilization is often "external," in the shed cuticle.*

Parthenogenesis appears to be lacking.

Asexual reproduction apparently does not occur; even anabiosis is lacking in aquatic forms.*

XXIX. PHYLUM PENTASTOMIDA

Habits — The tongue worms are all parasitic in vertebrate air passages. The larvae are also parasitic, usually in other vertebrates. Colonies are not formed.

Reproduction — Basic Bisexual.

Reproductive cycle — The parasitic adults produce gametes which fuse to form a zygote, which develops into a larva. The larva may molt several times and is surrounded by a cyst in the tissues of the secondary host; it metamorphoses in the cyst and matures into an adult when taken into the final host. The larva recalls the larva of some mites and also resembles some tardigrades. Little detail is available on the life cycle or the embryology. The species are dioecious and fertilization internal.*

Basic Bisexual Reproduction is the only known method; the species are dioecious.*

Apomictic reproduction is not known.

XXX. PHYLUM ONYCHOPHORA

Habits — These myriapod-like animals are terrestrial in the tropics. (A supposed marine fossil of Cambrian age is probably not an onychophoran.) They are solitary and never parasitic.

Reproduction — Basic Bisexual.

Reproductive cycle — The dioecious adults produce gametes which fuse to form a zygote. The spermatozoa are enclosed in a spermatophore. The species are oviparous (rarely), ovoviviparous, or viviparous. Embryonic development is usually discoidal — similar to that of arthropods. The larvae usually are only briefly mentioned, but at hatching they resemble the adult closely.* Cleavage may be superficial or total, with much yolk or none. The embryos may obtain nourishment from the mother by a placenta-like device in the uterus, or be enclosed in a tough shell. The eggs may be deposited before hatching, or the larva may develop to adult form before emergence.*

Basic Bisexual Reproduction no doubt is the only method used, although copulation has never been observed. The species are all dioecious, males are always present, and mature females always contain spermatozoa.*

Apomictic reproduction is not known.

XXXI. PHYLUM ARTHROPODA

Habits — The members of this largest phylum may be microscopic or large (6 ft long in some fossils). They may be marine, fresh-water, terrestrial, or virtually aerial. Many are parasitic, and a few form societies which are sometimes called "colonies." Some of the fresh-water species are not truly aquatic, as they breathe air. In many the larvae are the aquatic or parasitic forms. Parasitism is most often really micropredation, the host being killed automatically.

Reproduction — Basic Bisexual. Self-fertilization; gonad transplant. "Paedogenesis." Parthenogenesis — obligate, cyclic, or occasional. Asexual (budding, stolonization, polyembryony).

A. Class Merostomata

Reproductive cycle — The species are all dioecious. Adults in pairs produce gametes, which fuse to form a zygote, which develops into a "trilobite" larva. The larva is free-swimming; it molts repeatedly to change gradually into the adult form.

Basic Bisexual Reproduction is universal and exclusive, as all species are dioecious and fertilization occurs externally at the time of oviposition, by a single male.*

Apomictic reproduction is unknown.

B. Class Pycnogonida

Reproductive cycle — The species are dioecious, and the adults in pairs produce gametes which fuse to form a zygote. The eggs are usually brooded on the legs of the male. A protonymphon larva hatches; it later metamorphoses into an adult.

Basic Bisexual Reproduction is universal among these dioecious animals, as the male fertilizes the eggs as they are laid and then broods them.*

Apomictic reproduction is not known.

C. Class Arachnida

Reproductive cycle — The dioecious adults in pairs produce gametes, the sperm generally in spermatophores, except in some Acarina. The zygote develops either by meroblastic cleavage or total cleavage, either within the ovaries or brooded on the body or in a cocoon or in the ground. The young that hatch are juveniles and molt repeatedly to become adults. In the Ricinulei and Acarina a six–legged "larva" molts to the adult form and eight legs, sometimes passing through a protonymph stage, followed by deutonymph and tritonymph stages. In the ticks, stages are called larva, nymph, and adult. In some mites the larva may be followed by a hypopus stage, designed for attachment to a host or movement by wind.

Basic Bisexual Reproduction is nearly universal, as all species are dioecious* and parthenogenesis is known in only a few species.* Spermatophores are produced in several groups.

Parthenogenesis occurs in the cycle of several families of Acarina and in a harvestman, *Phalangium opilio.**

Asexual reproduction is not known.

D. Class Crustacea

Reproductive cycle — Adults may be hermaphroditic, although the species are usually dioecious; they usually copulate and produce gametes which fuse to form a zygote. From the egg hatches a larva, sometimes a juvenile but often very different from the adult. In the latter cases, the early stage in some species has been called a nauplius larva; it develops into a metanauplius, then into a protozoea, then into a zoea, and finally into an adult. These changes take place at molts. In still others the nauplius appears first and develops through a cypris larva. Many variations occur, but metamorphosis is always gradual.

Basic Bisexual Reproduction unquestionably occurs, but a variety of apomictic sexual processes occur also. The species are generally dioecious, usually not self-fertilizing, and usually not parthenogenetic. Spermatophores are sometimes produced.

Self-fertilization — Hermaphroditism occurs occasionally in most orders and predominantly among the Cirripedia,* in certain parasitic Isopoda, and in exceptional species in other groups.* There is some difference of opinion as to whether self-fertilization occurs, but it does occur in the parasitic Rhizocephala* and probably in other orders rarely. In the rhizocephalid *Peltogasterella,* a male cyprid larva extrudes dedifferentiated cells into the female; these redifferentiate into a testis, thus converting the female into an hermaphrodite. The eggs, therefore, are fertilized amphimictically.

Parthenogenesis is common in Branchiopoda and Ostracoda, where males are unknown in some species.* It also occurs in such barnacles as *Scalpellum, Ibla,* and

Trypetesa.* Alternation of parthenogenesis with bisexual reproduction occurs in the Cladocera.* Gonomery is said to occur.*

Budding and stolonization are reported in the Rhizocephala, but details are not available.*

Paedogenesis, of unspecified nature, is reported in Cladocera and Cirripedia.*

E. Class Pauropoda

Reproductive cycle — The species are always dioecious; the adults produce gametes which fuse to form a zygote. The eggs hatch into a larva with three pairs of legs (whereas the adult has ten pairs) followed by three other larval stages.

Basic Bisexual Reproduction is to be assumed, because the species are dioecious,* but no details are available.

Parthenogenesis is not reported but cannot be excluded.

Apomictic reproduction is not known.

F. Class Symphyla

Reproductive cycle — The species are all dioecious; the adults produce gametes, with the spermatozoa in a spermatophore, which fuse to form a zygote. The larva that hatches has 6 or 7 pairs of legs (the adult has 12 pairs) and adds segments at each molt.

Basic Bisexual Reproduction occurs, as the species are dioecious, and spermatozoa are transferred by spermatophore.* (The male deposits the spermatophore on a stalk, whence it is taken into the mouth of the female; the spermatozoa are then stored in special buccal pouches. The female removes an egg individually from her gonopore with her mouth, inseminates it with spermatozoa, and attaches it to a substratum.*

Parthenogenesis is common.*

Asexual reproduction is unknown.

G. Class Diplopoda

Reproductive cycle — The species are dioecious, and the adults produce gametes which fuse to form a zygote. Spermatozoan transfer is indirect, through a modified pair of legs or a spermatophore. The larva usually has 3 pairs of legs (the adult may have more than 50), and segments are added at each molt.

Basic Bisexual Reproduction is the rule, the species being dioecious, and copulation or indirect spermatozoan transfer is practised.*

Parthenogenesis, as a possibility, cannot be excluded.

Asexual reproduction is unknown.

H. Class Chilopoda

Reproductive cycle — The species are dioecious, and the adults produce gametes which fuse to form a zygote. Spermatozoan transfer is by spermatophore. In some, the young at hatching have the full complement of segments and legs; in others they may have 4 pairs, which increase to 13 or so during molts, in what is considered metamorphosis of the insectan type.

Basic Bisexual Reproduction is apparently universal. The species are dioecious, and spermatozoa are transferred by spermatophore.*

Parthenogenesis is not reported but also not excluded.

Asexual reproduction is unknown.

I. Class Insecta

Reproductive cycle — The species are nearly all dioecious, but a few are hermaphroditic and many parthenogenetic. The adults copulate and produce gametes which fuse to form a zygote. Direct insemination by means of genitalia on the apex of the abdomen occurs in most, but in Odonata the spermatozoa are transferred to special copulatory appendages at the base of the abdomen; and spermatophores are used in several orders. (These are placed in the female directly, not attached outside for pick-up.) The egg hatches into a nymph which molts to assume adult form gradually, or into a "larva" which changes into a pupa before reaching maturity. Most insects are oviparous, but viviparity occurs in several orders, often associated with parthenogenesis or paedogenesis. The usual cycle may be interspersed with parthenogenetic generations, paedogenesis, or polyembryony.

Basic Bisexual Reproduction surely predominates, as virtually all species are dioecious, but many variations occur that are apomictic. Polyspermy is common.* Copulation is nearly universal.

Self-fertilization — Hermaphroditism apparently occurs in the coccid *Icerya*,* where a single gonad produces both eggs and spermatozoa, with self-fertilization the normal condition. (Males occasionally occur and can fertilize the eggs.) In some stoneflies, there may be an ovary in the male, but it is nonfunctional.* Hermaphroditism also occurs in *Pediculus*.*

Parthenogenesis may be apomictic, automictic, or haploid.* It may alternate with amphigonous phases (called heterogamy by Hagen), as in Aphididae and Cynipidae.* It is also described as sporadic, constant, or cyclic. It may give rise to males, females, or both. It is obligate in the beetle *Graphognathus leucoloma** and is also known in one species of Embioptera* and in members of at least 11 other orders.

Polyembryony occurs in Strepsiptera,* many Hymenoptera,* rarely in the Acrididae (Orthoptera), and probably others. It may occur at the first cleavage, by division of a morula or a blastula, or by more complicated processes.*

Paedogenesis occurs in larvae of the cecidomyiid fly *Miastor*, by parthenogenesis,* and in *Micromalthus* (Coleoptera).* It occurs in pupae of *Chironomus grimmii* (Diptera).* In these latter it is simply precocious sexuality.

XXXII.   PHYLUM CHAETOGNATHA

Habits — The arrow worms are all small marine animals, usually planktonic but in one genus benthonic. They are always predaceous and do not form colonies.

Reproduction — Basic Bisexual. Self-fertilization occurs.

Reproductive cycle — The hermaphroditic adults produce gametes which fuse to form a zygote. The spermatozoa are transferred in a spermatophore, sometimes mutually, or self-fertilization occurs. The egg hatches into a minute larva, not very different from the adult, which develops directly.

Basic Bisexual Reproduction apparently occurs in most of the few genera except for *Sagitta*. They are all hermaphroditic, and in most the spermatozoa are transferred directly, so that cross-fertilization is probably predominant.

Self-fertilization — Hermaphroditism is universal,* usually protandric, but spermatozoa are stored in the body after transfer by spermatophore.* Self-fertilization is apparently the rule in *Sagitta*.*

Asexual reproduction is unknown, even though "surprising ability to regenerate" is reported.*

XXXIII.   PHYLUM POGONOPHORA

Habits — The beard worms are excessively long but slender worms, always solitary and living in secreted tubes. They are all marine, ranging from littoral waters to abyssal depths. None are parasitic nor colonial.

Reproduction — Basic Bisexual. Parthenogenesis is possible.

Reproductive cycle — Very little is known for sure of the life cycle of these abyssal animals. The sexes are separate, and apparently fertilization is in the tube of the female. Spermatophores may be used, consisting of a ball that is likened to a "morula." Gametes are produced, which fuse to form a zygote. The egg hatches into a larva with a girdle of cilia. Brooding occurs in the tube of the female, and the larva probably leads a brief free existence.

Basic Bisexual Reproduction appears to be the only process. The species are dioecious, so far as known.*

Parthenogenesis — The young are brooded, and parthenogenesis cannot be excluded as a possibility.

Asexual reproduction is unknown.

XXXIV. PHYLUM ECHINODERMATA

Habits — Echinoderms are exclusively marine, living on the bottom at all depths, with a few floating forms. Their larvae are free-swimming. They do not form colonies. Some are ectocommensal, but none are truly parasitic. They are never microscopic nor very massive.

Reproduction — Basic Bisexual. Self-fertilization sometimes is possible. Parthenogenesis. Asexual (division, fragmentation, "budding," polyembryony said to be sporadic).*

A. Class Crinoidea

Reproductive cycle — The species are dioecious; the adults produce gametes which fuse to form a zygote. The eggs are either liberated to be fertilized in the sea, or only the spermatozoa are liberated, to find their way into the brood pouch where the eggs are waiting. The egg hatches into a doliolaria larva, which attaches to the bottom (even in pelagic species) and develops into a pentacrinoid larva. It then matures into a stalked adult or breaks away from the stalk in pelagic species.

Basic Bisexual Reproduction is apparently universal. The species are all dioecious,* and fertilization is usually external in a marsupium.* In *Isometra vivipara,* spermatozoa are present in the genital tube of the ovary, so that the eggs are apparently fertilized before extrusion.*

Apomictic reproduction does not occur, although parthenogenesis is possible. Autotomized arms may be regenerated, but the arms always die.*

B. Class Somasteroidea

Until 1961 this group was known only from fossils of relatively poor preservation. It was then discovered that the genus *Platasterias,* known from the Pacific Ocean off Mexico since 1871, may be a somasteroid. Little is recorded about this genus directly, but it was for long listed as an asteroid and may be presumed to share features of echinoderms in general and probably also of asteroids.

C. Class Asteroidea

Reproductive cycle — All species are dioecious, but hermaphroditic individuals may occur. Adults produce gametes which fuse to form a zygote, with both egg and spermatozoon liberated into the sea, usually simultaneously by a pair or a group. A variety of brooding methods are used. There may be either indirect development from the egg, through a free-swimming ciliated coeloblastula, a bipinnaria larva, and a brachiolaria larva, to an adult, or direct development in which one or both of the larval stages is omitted. The maturing from the last larval stage is cited as a metamorphosis. This basic cycle may be interrupted by division through the disk or autotomy of an arm with regeneration of both parts.

Basic Bisexual Reproduction is predominant, but hermaphroditic specimens occasionlly occur in many species, sometimes involving most of a population.* Sex reversal occurs.*

Self-fertilization — Hermaphroditism seems to be limited to certain individuals or populations. Sometimes the gonads are united, and protandry does not seem to be universal,* so self-fertilization cannot always be excluded.

Division ("fission") through the disk occurs in several families, with regeneration of both parts.* This process is sometimes called schizogamy. It may also occur in juveniles.* Autotomized arms are regenerated into new individuals only in *Linckia*.*

D. Class Ophiuroidea

Reproductive cycle — The species may be dioecious or hermaphroditic; the adults produce gametes which fuse to form a zygote, usually with both eggs and spermatozoa liberated into the sea, occasionally with the eggs cemented to rocks. The eggs may be brooded, in special bursae or even in the ovary, where fertilization must then take place. The egg may undergo an indirect development, with an ophiopluteus larva that metamorphoses into a juvenile; or in brooding species it may develop without the larval stage, sometimes becoming a juvenile before liberation from the body of the mother. The basic cycle may be interrupted by division.

Basic Bisexual Reproduction occurs in the majority, which are dioecious.*

Self-fertilization — Hermaphroditism occurs in many species of Ophiurae such as *Amphipholis squamata*.* Some are protandric, some apparently change sex several times,* and in some there is an ovotestis or two gonads simultaneously.* All monoecious species brood their young.* When eggs and spermatozoa are shed into surrounding water, self-fertilization can occur by accident in some species but apparently is never obligate.* Anomalous hermaphroditic individuals occur in normally dioecious species.*

Parthenogenesis was reliably reported in three antarctic species, including *Ophiacantha vivipara*, in which only females were found.*

Division ("fission") occurs through the disk in some species with small six-armed individuals, especially species of *Ophiactis*. This is not clearly paratomous but seems to be a form of autotomy rather than of fragmentation.*

E. Class Echinoidea

Reproductive cycle — The species are mostly dioecious, with rarely an hermaphroditic individual; a few species are fully monoecious. Adults produce gametes which fuse to form a zygote, and both ova and spermatozoa are commonly shed into the sea. The eggs may be brooded in depressions of the outer surface. The egg develops into a ciliated free-swimming blastula, which gastrulates and becomes a pluteus (or echinopluteus) larva. This metamorphoses into a young urchin.

Basic Bisexual Reproduction is the only method reported but hermaphroditic individuals may occur rarely and there are species with combined gonad (ovotestis).*

Self-fertilization — Hermaphroditism occurs only as anomalous individuals in otherwise dioecious species. In these cases the sex cells are commonly self-fertile, so self-fertilization may occur,* although this would be external.

Parthenogenesis is fairly well substantiated in several species* and can be produced artificially in others.

Asexual reproduction is not known.

F. Class Holothurioidea

Reproductive cycle — Adults produce gametes which fuse to form a zygote, the ova and spermatozoa being shed into the sea (in oviparous species) or only the spermatozoa (in viviparous or brooding species). The egg develops into a free-swimming auricularia larva. This transforms into a doliolaria larva, which becomes a pentactula larva and then an adult. One or all of these larval stages may be omitted, in the latter case the brooded embryos appear first as juvenile holothurians. Fragmentation is the only process that interrupts this basic cycle.

Basic Bisexual Reproduction surely predominates, as the majority of species are dioecious,* but other processes cannot be excluded.

Self-fertilization — Hermaphroditism occurs in Dendrochirota such as *Cucumaria* and in some Synaptida,* and in several cases eggs and spermatozoa are produced simultaneously in each gonadial tubule. They are shed at separate times but seem to be self-fertile, but self-fertilization is not reported.*

Parthenogenesis is not reported but cannot be excluded as a possibility.

Fragmentation by transverse constriction occurs in many species, either dioecious or monoecious.* It may be multiple.* It is claimed that fragments thrown back into the sea will regenerate; and it was demonstrated by Deichmann in 1921 that "autotomy" is common in some species.

Budding has been reported but without details.*

XXXV. PHYLUM PTEROBRANCHIA

Habits — These are small or microscopic animals, all but one of which (*Atubaria heterolopha*) consist of colonies of many individuals.* All are marine, feeding on detritus or plankton. None are parasitic.

Reproduction — Self-fertilization is possible. Parthenogenesis is likely. Asexual (budding).

Reproductive cycle — This phylum consists of three genera sufficiently different that they are seldom described together. Most of the cycles are little known. The following is the available knowledge for the three genera:

Cephalodiscus — It lives in a colony of disconnected individuals in which the members of the colony coenoecium may be of one sex or two; individuals may also be hermaphroditic. The zooids produce gametes, but the site of fertilization is not clear. The eggs are shed into spaces in the coenoecium. The embryo at the gastrula stage becomes ciliated and swims in the sea. It changes gradually into an adult. Budding produces new individuals outside of this cycle.

Atubaria — It lives singly with no covering. The zooids are similar to those of *Cephalodiscus,* but only females and juveniles are known, and nothing is reported on the life cycle. Budding is not known.

Rhabdopleura — It lives in a branching stolonic colony, in which the zooids remain in contact with each other. Most known colonies are sterile, but a few have male and female zooids among the neuter ones. The site of fertilization seems to be unknown, but the eggs are shed into the sea. Later development is not reported.

Basic Bisexual Reproduction does not occur, because budding produces the colonies, which are thus clonal.

Self-fertilization — Hermaphroditism is not uncommon, but it is not clear whether egg and spermatozoa are produced simultaneously. In *Cephalodiscus sibogae,* the only known colony consisted of male and neuter zooids only.

Parthenogenesis may occur in *Atubaria,* in which only females are known.

Budding produces the colonies; the buds arise from the stalk in *Cephalodiscus* or from stolons in *Rhabdopleura.*

XXXVI. PHYLUM ENTEROPNEUSTA

Habits — The acorn worms are solitary and worm-like. They are marine and occur in fairly shallow water. They generally live in burrows, are always solitary, and never form colonies. The larvae are planktonic. None are parasitic.

Reproduction — It is Basic Bisexual. Asexual (fragmentation).

Reproductive cycle — The species are dioecious, and the adults produce gametes that fuse to form a zygote, which develops into a tornaria larva. The larva develops gradually into an adult. The gametes are released into the sea, where fertilization takes place. In some species development is direct, bypassing the tornaria larva stage. There is usually no asexual reproduction.

Basic Bisexual Reproduction, with the species dioecious, seems to be universal.

Fragmentation occurs in *Balanoglossus capensis,* where small pieces are cut off from the tail end forward, with each regenerating completely.

XXXVII. PHYLUM PLANCTOSPHAEROIDEA

Habits — The only known species is a larva or a larviform adult. The animals are ciliated and free-swimming in the sea. They are neither colonial nor parasitic. They are presumed to be the larvae of an unknown animal something like an enteropneust.

Reproduction: Gonads are not found, and no form of reproduction has been reported.

XXXVIII. PHYLUM TUNICATA

Habits — The tunicates are all marine, either attached to the bottom or planktonic. They are either solitary or colonial, and they are never parasitic.

Reproduction — It is Basic Bisexual. Self-fertilization is possible, clonal fertilization. Parthenogenesis is possible. Asexual (fragmentation, budding, strobilation). Colonial (detachment of individuals).

A. Class Larvacea (Appendicularia)

Reproductive cycle — These free-swimming animals are believed to be neotenic, to have lost the adult stage. It is sometimes hinted that the individuals are self-sterile, but gametes are produced, which fuse to form a zygote. This develops into a tadpole larva which undergoes some metamorphosis without assuming an adult form.

Basic Bisexual Reproduction presumably occurs. The individuals are generally protandrous hermaphrodites. *Oikopleura dioica* is dioecious.

Self-fertilization cannot be excluded, as hermaphroditism is nearly universal.

Parthenogenesis is possible although unreported.

Asexual reproduction is not known.

B. Class Ascidiacea

Reproductive cycle — The species are mostly hermaphroditic, but dioecious ones occur. Adults produce gametes which are liberated. The zygote is in the sea or may be brooded. In the latter case, all development occurs in the brood chamber, but in free-living forms a tadpole or appendicularia larva is liberated, which metamorphoses into an adult on the bottom. Budding and fragmentation, in both larva and adult, may interrupt this cycle.

Basic Bisexual Reproduction probably does occur rarely in solitary species. Most species are dioecious. Even in hermaphroditic species, it is reported that spermatozoa will not unite with ova from the same individual.

Self-fertilization cannot be excluded, because of hermaphroditism, but it is not reported.

Fragmentation occurs in the new buds of *Distaplia* immediately after detachment; each of the four or five parts grows into a new individual. In the Polyclinidae, soon after attachment of the larva, a long postabdomen divides into a string of new zooids. In *Diazona,* eight or nine spherical fragments split from a regressed zooid with the advent of winter.

Budding is diverse but occurs only in colonial forms. Buds may form from the mantle (pallial budding), the thoracic wall, the oesophagus, the water ampullae, or from stolons, some of which may develop only after a prolonged interval (resting buds). In pyloric budding in *Trididemnum,* one bud forms on the oesophagus and a second from the epicardial tubes; together the two buds produce a new individual. Pallial budding may start in the larval stage, and pyloric budding (double, as described above) may originate in an embryonic stage.

Strobilation is reported in Diazonidae, but details are lacking.

Colonial duplication — Individuals may detach from a colony and give rise to new colonies.

C. Class Thaliacea

Reproductive cycle — The hermaphroditic adult zooids produce gametes which fuse to form a zygote, and fertilization can usually be presumed to be free in the sea, although in some solitary ones it is internal. Budding produces the colonies. Examples are:

Pyrosoma — The egg is retained in the parent and becomes an oozooid, which produces a stolon with four buds. The buds are liberated, the oozooid dies, and the four buds together form a colony by secondary budding. The colony becomes pelagic.

In solitary relatives of *Pyrosoma,* the chain of buds separates to form sexually reproducing adults.

Doliolum — The egg is shed into the sea and develops into a tailed larva. This larva metamorphoses into an adult zooid, which produces a stolon with buds (prebuds). These give rise to several generations of specialized buds which remain attached. Eventually some buds develop into sexually reproducing adults, which detach and swim away.

Salpa — The fertilized egg develops into an oozooid, which buds a stolon with additional buds, which become blastozooids; the latter becomes the new hermaphroditic sexual individuals.

Basic Bisexual Reproduction presumably predominates or is universal.* All are hermaphrodites, apparently.*

Self-fertilization is not reported but cannot be excluded in hermaphrodites. In the Salpida, fertilization is generally by sperm from another individual in the same chain and therefore clonal.*

Fragmentation occurs in the Pyrosomata, when stolons divide into fragments which give rise to new individuals. This occurs early in the growth of the primary zooid.

Budding from a stolon produces zooids of several types, resulting in a colony.*

XXXIX. PHYLUM CEPHALOCHORDATA

Habits — The lancelets are marine animals of fish-like form, free-swimming but often buried in bottom sand. They are filter-feeders. None are colonial or parasitic.

Reproduction — It is Basic Bisexual. Self-fertilization is possible.

Reproductive cycle — Adults produce gametes which fuse to form a zygote. Fertilization is in the sea. A free-swimming ciliated larva is hatched and gradually assumes adult form.

Basic Bisexual Reproduction presumably occurs, as most species are dioecious.*

Self-fertilization — Hermaphroditism is said to occur in a number of cases,* but these are not listed nor described in available literature.

Asexual reproduction is not known.

XL. PHYLUM VERTEBRATA

Habits — The vertebrates are small to very large animals, marine, fresh-water, or terrestrial. They never are colonial, except in a social sense. Parasitism occurs occasionally as ectoparasites (hagfishes).

Reproduction — It is Basic Bisexual. Self-fertilization is possible. Parthenogenesis is rare. Asexual (polyembryony in armadillos and sporadically in all classes).

A. Class Agnatha

Reproductive cycle — Adults in pairs produce gametes which fuse to form a zygote. Fertilization is external. In some an ammocoetes larva is developed, which metamorphoses into an adult. In others, development is direct.

Basic Bisexual Reproduction is nearly universal, with fertilization external.*

Self-fertilization — Hermaphroditism is both reported and denied in *Myxine glutinosa*, but it is believed quite unlikely that it is self-fertilizing.* Oocytes and spermatocytes occur together in the gonad of young lampreys, before sexual differentiation occurs, but apparently there is no fertilization at that time.*

Asexual reproduction is not known.

B. Class Chondrichthyes

Reproductive cycle — The species are all dioecious; the adults produce gametes which fuse to form a zygote. Fertilization is internal. Development is direct.

Basic Bisexual Reproduction occurs exclusively in all species, so far as known.* The species are dioecious, and fertilization is internal.* But gonomery is reported in *Raja*.*

Hermaphroditism is reported only as an abnormality.*

Apomictic reproduction is not known, although parthenogenesis cannot be excluded.

C. Class Osteichthyes

Reproductive cycle — Most species are oviparous, but some are viviparous. Adults produce gametes which fuse to form a zygote. Usually development is direct, but sometimes a "larva" intervenes which is different from the adult, but development is still gradual.

Basic Bisexual Reproduction predominates, with the species usually dioecious and fertilization external.*

Self-fertilization — Hermaphroditism occurs in several families. In the Symbranchidae it is protogynous.* In the Sparidae and Maenidae it is said to be well-developed.* In the Serranidae there is said to be some "functional hermaphroditism," by which is apparently meant self-fertilization.*

Parthenogenesis has been reported, but it is not clear whether normal development ever occurs.* Gonomery occurs in *Fundulus*.*

Asexual reproduction is not known.

D. Class Amphibia

Reproductive cycle — The species are all dioecious. Adults produce gametes which fuse to form a zygote. Fertilization is external or internal. Sometimes the spermatozoa are enclosed in a gelatinous spermatophore.* An aquatic larva (tadpole) usually occurs metamorphosing to adult form. Neoteny occurs in some salamanders, where larvae reach adult size and even breed while retaining larval features.

Basic Bisexual Reproduction predominates, with the sexes separate and fertilization external.*

Hermaphroditism is reported only as an abnormality.*

Parthenogenesis has been reported in salamanders.*

Polyembryony is said to be sporadic.*

Paedogenesis, sexual, occurs normally in the Mexican axolotl,* but the fact that this animal never attains adult form makes it more appropriate to call this neoteny.

E. Class Reptilia

Reproductive cycle — The species are nearly all dioecious. Adults produce gametes which fuse to form a zygote. As most species are oviparous, the egg develops outside, hatching into a juvenile. A few species are viviparous or ovoviviparous and bear living juveniles.

Basic Bisexual Reproduction predominates, with most species being dioecious and fertilization being internal.*

Self-fertilization – Hermaphroditism is known in Chelonia, with an ovotestis,* but self-fertilization is not reported.*

Parthenogenesis is the only process in certain races of *Lacerta saccicola* in the Caucasus* and some species of *Cnemidophorus* in America.*

Polyembryony is said to occur sporadically.*

F. Class Aves

Reproductive cycle — The dioecious adults produce gametes which fuse to form a zygote. Fertilization is always internal. The eggs are laid and hatch into a juvenile bird.

Basic Bisexual Reproduction is universal, with fertilization internal after copulation.*

Hermaphroditism is reported only as an abnormality.*

Parthenogenesis has apparently occurred in artificial breeds of domestic fowl.*

Polyembryony is said to occur sporadically.* Double-yolked eggs may be the result.

G. Class Mammalia

Reproductive cycle — All are dioecious, and except for monotremes are all viviparous. Adults produce gametes which fuse to form a zygote, and fertilization is always internal. The armadillos lay eggs, from which juveniles hatch. Their cycle is always altered by polyembryony, which is occasional in other mammals also (identical twins). Marsupials bear their young at such an early embryonic stage that they are almost more embryos than juveniles.

Basic Bisexual Reproduction is nearly universal, with fertilization internal after copulation.

Hermaphroditism is reported only as an abnormality.*

Parthenogenesis has been reported, but it appears never to result in successful development.*

Polyembryony occurs in armadillos, in the blastocyte stage.* In humans identical twins arise from polyembryony; this process is sporadic in the class.*

Chapter 7

DISTRIBUTION OF PROCESSES

Activation — In the restricted sense, occurs in some Protozoa, Turbellaria, Nematoda and Gastropoda. In the broad sense, also includes all cases of Parthenogenesis and Syngamy.
Agametes — See Agamogony.
Agamogenesis — See Asexual Reproduction.
Agamogony — Occurs in Protozoa (Sporozoa) and Mesozoa (Dicyemida and Orthonectida).
Agamy I (nongametic unisexual) — Occurs in Turbellaria, Bdelloidea, Monogononta, Chaetonotoidea, Nematoda, Gastropoda, Sipunculoidea, Heterotardigrada, Crustacea, Insecta, Ophiuroidea, and Pterobranchia.
Agamy II (without gametes) — See Asexual Reproduction.
Agamy III (reproduction by agametes) — Occurs in Protozoa and Mesozoa.
Allogamy — See Outbreeding.
Ameiotic Reproduction — Occurs in some Protozoa, Mesozoa, Rotifera, Nematoda, Crustacea, Insecta, and all those listed under Asexual Reproduction. See also Parthenogenesis (Ameiotic).
Amixis I — In the strict sense, would include all reproduction that does not involve Karyogamy.
Amixis II (without mixing) — As the absence of mixing, includes all under Parthenogenesis and Asexual Reproduction.
Amphigenesis — Includes all under Bisexual Reproduction and Hologamy.
Amphigony I (sexual reproduction) — Same as Sexual Reproduction.
Amphigony II (requiring two individuals) — Same as Amphigenesis.
Amphimixis — Occurs in all gonochoristic or dioecious animals, when there is karyogamy.
Amphitoky — Same as Deuterotoky. Term used only in Insecta.
Androgenesis — Same as Arrhenotoky. Term used principally in Insecta.
Anisogametes — See Anisogamy.
Anisogamy — Occurs in all dioecious species except those Protozoa using Isogametes or Hologametes.
Anthogenesis — Same as Deuterotoky. Term used only in Insecta.
Apomixis — Includes Parthenogenesis, Automixis, and all Asexual Reproduction.
    No Apomixis — Temnocephaloidea, Cestodaria, Acanthocephala, Rotifera, Priapuloidea, Gordioidea, Echiuroidea, Myzostomida, Pentastomida, Onychophora, Merostomata, Pycnogonida, Pauropoda, Diplopoda, Chilopoda, Crinoidea, Planctosphaeroidea, and Chondrichthyes.
Architomy — Occurs in Mesozoa, Hydrozoa, Scyphozoa, Anthozoa, Ctenophora, Turbellaria, Rhynchocoela, Polychaeta, Oligochaeta, Asteroidea, Ophiuroidea, Holothurioidea, Enteropneusta, Ascidiacea, Thaliacea, and probably others.
Arrhenotoky — Parthenogenesis producing only males. Occurs in Insecta.
Asexual Reproduction — Occurs in most phyla and most classes.
    No Asexual Reproduction — Acanthocephala, Rotifera, Gastrotricha, Kinorhyncha, Nematoda, Gordioidea, Priapuloidea, and Tardigrada, in all of which there is eutely. Also not reported in Brachiopoda, Mollusca, Echiuroidea, Myzostomida, Pentastomida, Hirudinea, Archiannelida, Dinophiloidea, Onychophora, Arachnida, Symphyla, Diplopoda, Chilopoda, Chaetognatha,

Pogonophora, Echinoidea, Larvacea, Cephalochordata, Agnatha, and Chondrichthyes, where polyembryony may nevertheless occur.

>Only asexual — A few species in Sarcodina, Turbellaria, Phylactolaemata, Oligochaeta, and probably others.

Assembled Bodies — Occur in Porifera (gemmules and reduction bodies), Scyphozoa (podocysts), and Bryozoa (statoblasts and hibernacula).

Autogamy I (fertilization by sister nuclei) — Occurs only in Protozoa (Ciliata).

Autogamy II (self-fertilization) — Occurs in Protozoa (Sarcodina and Sporozoa).

Automixis — Includes autogamy and paedogamy. Occurs only in Protozoa.

Autotomy (with regeneration of both parts) — Occurs in Turbellaria, Rhynchocoela, Asteroidea, Holothurioidea, and probably others.

Basic Bisexual Reproduction — Occurs in many groups, but not in any species which includes in its life cycle "any" of the following: Autogamy, Hologamy, or Apomixis or Asexual Reproduction of any sort. It seems to be the only method in Gnathostomuloidea, Acanthocephala, Seisonidea, Kinorhyncha, Priapuloidea, Gordioidea, Brachiopoda, Scaphopoda, Cephalopoda, Echiuroidea, Myzostomida, Eutardigrada, Pentastomida, Onychophora, Merostomata, Pycnogonida, Pauropoda, Diplopoda, Chilopoda, Pogonophora, Crinoidea, Enteropheusta, Chondrichthyes, and Aves.

>No Basic Bisexual Reproduction — Protozoa (Sarcodina, Flagellata, Ciliata, Suctoria, Sporozoa), Mesozoa, Monoblastozoa, Coelenterata (unlikely), Bdelloidea, Calyssozoa, Bryozoa, and Pterobranchia. In many other groups it is rare.

Binary Fission — Occurs only in Protozoa and Mesozoa (agametes), by definition (must be unicellular).

Bisexual Reproduction — It is absent in many individuals and in widely scattered species (i.e., some Hydrozoa, some Rhynchocoela, some Polychaeta, and some Oligochaeta). It is entirely absent from all species in the Chrysomonadina (Protozoa), the Bdelloidea (Rotifera), and the Chaetonotoidea (Gastrotricha). So far as known, it is the only reproductive method in these groups: Gnathostomuloidea, Cestodaria, Acanthocephala, Seisonidea, Kinorhyncha, Priapuloidea, Gordioidea, Brachiopoda, Scaphopoda, Cephalopoda, Echiuroidea, Myzostomida, Eutardigrada, Pentastomida, Onychophora, Merostomata, Pycnogonida, Pauropoda, Diplopoda, Chilopoda, Pogonophora, Crinoidea, Enteropneusta, Cephalochordata, Chondrichthyes, and Aves. It is the only known method in some species in many other groups, including Porifera, Coelenterata, Ctenophora, Platyhelminthes, Rhynchocoela, Monogononta, Macrodasyoidea, Nematoda, Phoronida, Mollusca, Annelida, Arachnida, Crustacea, Insecta, Chaetognatha, Asteroidea, Ophiuroidea, Holothurioidea, Enteropneusta, Tunicata, and Vertebrata.

>No Bisexual Reproduction — Some Protozoa (Chrysomonadina), some Rotifera (Bdelloidea), some Gastrotricha (Chaetonotoidea), some Rhynchocoela *(Lineus sanguineus),* some Polychaeta *(Zeppelina),* some Oligochaeta (Naididae and *Lumbriculus variegatus*).

Budding — Occurs in Protozoa (Sarcodina, Suctoria, Ciliata), Porifera (Calcarea, Hexactinellida, Demospongia), Hydrozoa (polyp, medusa, and actinula larva), Scyphozoa (planula larva, scyphistoma larva, strobilus), Anthozoa, Cestoda (coenurus bladder, hydatid cyst, cysticercus, strobilus), (Calyssozoa embryo, stolon), Gymnolaemata, Phylactolaemata, Phoronida, Polychaeta (adult), Oligochaeta (adult), Crustacea, Holothurioidea, Pterobranchia (zooids), Ascidiacea (zooid, larva),

Thaliacea (stolon).
> Double Budding — Occurs in Hydrozoa.
> Multiple Budding — Occurs in Flagellata, Sarcodina, Phylactolaemata, and Tunicata.
> Pyloric Budding (dual) — Occurs only in Ascidiacea.

Clonal Fertilization — Probably occurs in most classes, where clones may go undetected; reported in Hydrozoa, Anthozoa, Trematoda, Calyssozoa, Phylactolaemata, Gymnolaemata, Pterobranchia, and Thaliacea.
Configuration — Occurs only in Protozoa (Ciliata and Suctoria).
Cross-fertilization — Occurs in most species in most classes, but is absent in all Chrysomonadina (Protozoa), Bdelloidea, Chaetonotoidea, and occasional species in many groups. (See also Outbreeding)
Cytogamy — Occurs only in Protozoa (Ciliata).

Deuterotoky — Term used only in Insecta for parthenogenesis producing both males and females.
Dichogamy — Occurs in all protandric or protogynic species.
Dichotomous Autotomy — Same as architomy.
Dioecism — Occurs in most species in all classes having any sexual processes, except for Gastrotricha, Gnathostomuloidea, Phylactolaemata, Oligochaeta, and Hirudinea.
Dissogeny (dissogony) — Occurs only in Ctenophora, possibly in both of the "classes" usually cited (Tentaculata and Nuda).
Division — Occurs at some stage in the cycle in Mesozoa, Monoblastozoa, Hydrozoa, Scyphozoa, Anthozoa, Ctenophora (possibly), Turbellaria, Cestoda (cysticercus), Rhynchocoela, Gymnolaemata, Phoronida, Polychaeta, Oligochaeta, Insecta, Asteroidea, Ophiuroidea, Enteropneusta, Ascidiacea, Thaliacea, Mammalia, and probably others. (See also Fission)

Endogamy — Inbreeding can occur accidentally in almost any cross-fertilizing species.
Endomixis — Occurs only in Protozoa (Ciliata).
Epigamy — Occurs only in Polychaeta.
Epitoky — Occurs only in Polychaeta.
Etheogenesis — Occurs only in plants.
Exogamy — Same as Outbreeding.

Fission — Occurs only in Protozoa and single cells of Metazoa (by definition). (See also Division)
Fissiparity — Same as Architomy.
Fragmentation — Occurs in Porifera, Orthonectida (plasmodium and larva), Hydrozoa (polyp), Scyphozoa (medusa or scyphistoma), Anthozoa (polyp), Tentaculata, Gnathostomuloidea, Turbellaria, Trematoda, Cestoda (plerocercoids), Rhynchocoela, Phoronida, Polychaeta, Oligochaeta, Asteroidea, Ophiuroidea, Holothurioidea, Enteropneusta, Ascidiacea (buds, larva, zooid), Thaliacea (zooid), as well as those listed under Division, Architomy, Polyembryony, and perhaps even Budding.
Fraternal Fertilization — Can occur accidentally in any cross-fertilizing species. Definitely occurs in Dinophiloidea.
Frustulation — Occurs only in Hydrozoa and Turbellaria.
Fusion — Besides syngamy, fusion occurs to form new individuals in Porifera (larvae), Hydrozoa (oocytes), and Trematoda (diporpa larvae). In Sporozoa the process is called Syzygy.

Gametic Reproduction — Occurs in all groups with Gamogony.
Gametogenesis — Occurs in all groups using Gamogony.
Gamogenesis — Same as Sexual Reproduction.
Gamogony — Occurs in all groups that reproduce sexually, except for those Protozoa using Hologamy; thus all Amphimixis and Parthenogenesis.
Geneagenesis — Same as Parthenogenesis.
Gemmation — Term used only in Insecta.
Gemmiparity — Term used only in Polychaeta.
Gemmulation — Occurs only in Porifera (gemmules). (See also Hibernacula, Podocysts, Sorites, Statoblasts)
Gonad Transplant — Occurs only in Crustacea.
Gonochorism — Occurs in most species in all classes having any sexual processes, except for the completely parthenogenetic ones and those that are all hermaphroditic. (See Hermaphroditism and Parthenogenesis)
Gonomery — Occurs in Turbellaria, Gastropoda, Crustacea, Chondrichthyes, Osteichthyes, and Amphibia.
Gynogenesis I (entry without karyogamy) — See Pseudogamy.
Gynogenesis II (only females produced) — See Thelytoky.

Hemimixis — Occurs only in Protozoa (Ciliata).
Hermaphroditism — Occurs in some species in each of the following phyla or classes (the word "all" indicating that this occurs in "all classes" of the phylum, not in all species): all Porifera, Mesozoa (Orthonectida), all Coelenterata, all Ctenophora, all Platyhelminthes, Rhynchocoela, Gastrotricha (Macrodasyoidea), Nematoda, Gordioidea, Calyssozoa, all Bryozoa, Phoronida, Brachiopoda (Inarticulata), all Mollusca except Monoplacophora and Scaphopoda, Sipunculoidea, Myzostomida, all Annelida, Crustacea, Insecta, Chaetognatha, Asteroidea, Ophiuroidea, Echinoidea (rarely), Holothurioidea, Pterobranchia, all Tunicata, Cephalochordata (possibly), and Vertebrata (Agnatha, Osteichthyes, Reptilia). The corresponding list of the phyla which do not ever show hermaphroditism includes: Acanthocephala, Rotifera, Kinorhyncha, Priapuloidea, Echiuroidea, Dinophiloidea, Tardigrada, Pentastomida, Onychophora, Pogonophora, Enteropneusta, as well as some classes of Mollusca, Arthropoda, Echinodermata, and Vertebrata.
Heterogamy — Same as Anisogamy.
Hibernacula — Occur in some Bryozoa (Phylactolaemata).
Hologamy — Occurs only in Protozoa (Sarcodina and Flagellata).
Hypogenesis — Same as Asexual Reproduction.

Inbreeding — Can occur in any species using syngamy, by accidental mixture from closely related individuals.
Isogamy — Occurs only in some Protozoa (Sarcodina and Flagellata).

Karyogamy — Occurs in all truly fertilizing animals; thus, in all Basic Bisexual Reproduction, as well as in all groups where diploidy of a zygote is achieved by union of a spermatozoon with an ovum.

Meiosis — By definition occurs in all animals in which there is any sexual reproduction of any sort.
Merogony I (fertilization of enucleated egg fragment) — Occurs in experimental situations only, in Rhynchocoela, Scaphopoda, Echinodermata, and Amphibia.
Merogony II (production of merozoites) — Occurs only in Protozoa (Sporozoa).

Mixis — Essentially the same as Out-breeding but could include all cases of Karyogamy.
Monoecism — Occurs in all groups listed under Hermaphroditism.
Monogony — Term used only in Annelida.
Monotomy — Can occur only in Protozoa and Mesozoa.
Multiple Fission — Occurs in Sarcodina (flagellated young or spores), Flagellata (flagellated young), Opalinida (gametes), Sporozoa (spores, merozoites, schizonts, sporozoites, gametes), or in Mesozoa (agametes).
Mychogamy — Same as Self-fertilization.

Nuclear Reorganization — Occurs only in Protozoa, as Hemimixis or Endomixis.

Oogamy — Occurs in all instances of anisogamy except the rare cases in which the ovum is no larger than the spermatozoon.
Oogenesis — Occurs in all female individuals that take part in Sexual Reproduction.
Outbreeding — Occurs in all known species of Kinorhyncha, Solenogastres, Gastropoda, and Agnatha, in most species of Coelenterata, Platyhelminthes, Rhynchocoela, Nematoda, Bivalvia, Annelida, Arthropoda, Echinodermata, and other Vertebrata, and in a few species of Protozoa, Gastrotricha, Rotifera, and Chaetognatha. (See also Cross-fertilization)
Ovotestis — Occurs in Rhynchocoela, Amphineura, Gaztropoda, Bivalvia, Insecta, Asteroidea, Ophiuroidea, Echinoidea, Holothurioidea, Agnatha, and Reptilia.

Paedogamy — Occurs only in Protozoa (Ciliata and Sporozoa).
Paedogenesis — Occurs in Hydrozoa, Scyphozoa, Ctenophora, Trematoda, Cestoda, Crustacea, Insecta, Ascidiacea, and Amphibia.
Palintomy — Same as Sporogony; only in Protozoa (Sarcodina, Sporozoa).
Paratomy — Occurs in all groups using Schizometamery (Polychaeta), Epitoky (Polychaeta), Budding, or Strobilation. Also cited in Oligochaeta.
Parthenogenesis — It is obligate and universal or widespread in these groups: Bdelloidea, Monogononta, Chaetonotoidea, Heterotardigrada, Symphyla, and Insecta. It probably occurs in most other classes, where its occurrence sporadically would easily go undetected.
    Facultative Parthenogenesis — In even one isolated species, it is likely to be unknown or buried in literature, but it does occur in Turbellaria, Nematoda, Gymnolaemata, Gastropoda, Sipunculoidea, Dinophiloidea, Oligochaeta, Arachnida, Crustacea, Ophiuroidea, Echinoidea, Pterobranchia, Osteichthyes, Amphibia, and Reptilia. It is possible also in many other groups.
    Ameiotic Parthenogenesis — Occurs at least in some Rotifera, Nematoda, Mollusca, Crustacea, and Insecta.
    Automictic Parthenogenesis — Same as Meiotic Parthenogenesis.
    Diploid Parthenogenesis — Same as Ameiotic Parthenogenesis.
    Haploid Parthenogenesis — Same as Meiotic Parthenogenesis.
    Hemizygoid Parthenogenesis — Same as Haploid Parthenogenesis.
    Meiotic Parthenogenesis — This is said to be less common than Ameiotic Parthenogenesis; it occurs at least in Turbellaria, Oligochaeta, Crustacea, and Insecta, and probably elsewhere.
    Zygoid Parthenogenesis — Same as Diploid Parthenogenesis.
Patrogenesis — Same as Merogony I.
Pedal Laceration — Occurs in Hydrozoa, Scyphozoa (scyphistoma larva), Anthozoa, and Ctenophora (creeping forms).

Plasmogamy — Same as Plasmogony.
Plasmogony — Occurs in Protozoa (Sarcodina) and Nematoda.
Plasmotomy — Occurs only in Protozoa (Sarcodina, Opalinida, Sporozoa).
Plastogamy — Occurs only in Mycetozoa (Protozoa — Sarcodina).
Podocysts — Occur only in Scyphozoa (scyphistoma larva).
Polar Body Fertilization — Occurs in Crustacea and Asteroidea, and experimentally in Mollusca and Annelida.
Polyembryony — Occurs in some form in Coelenterata, Trematoda, Cestoda, Gymnolaemata, Oligochaeta, Insecta, Mammalia, and very likely others. Said to be sporadic in all major phyla. A similar process occurs in Protozoa (Sarcodina). (See also Successional Polyembryony)
Pre-buds — Term used only in Tunicata.
Progenesis — Occurs in some Trematoda (metacercaria larva).
Protandry — Occurs in Anthozoa, Gnathostomuloidea (possibly), Rhynchocoela, Gastrotricha, Nematoda, Bryozoa (Gymnolaemata), Phoronida, Mollusca (Solenogastres, Gastropoda, Bivalvia), Myzostomida, Annelida, Chaetognatha, and Ophiuroidea.
Protogenesis — The same as Budding.
Protogyny — Occurs in Anthozoa, Gastrotricha, Calyssozoa, Bryozoa (Gymnolaemata), Mollusca (Bivalvia), and Osteichthyes.
Pseudogamy — Occurs in Protozoa, Turbellaria, Nematoda, and Oligochaeta.

Reduction Bodies — Occur in very different forms in Porifera and in Tardigrada.
Resting Buds — Only in Ascidiacea.

Schizogamy I (division into unlike individuals) — Occurs only in Polychaeta.
Schizogamy II (ordinary division) — Term used only in Asteroidea.
Schizogenesis — The same as Fission.
Schizogony — Occurs only in Protozoa (Sporozoa).
Schizometamery — Occurs only in Polychaeta.
Scissiparity — The same as Architomy.
Self-fertilization — Occurs in Porifera, Hydrozoa, Scyphozoa and Anthozoa possibly, Ctenophora, Turbellaria, Trematoda, Cestoda, Rhynchocoela, Nematoda, Calyssozoa (possibly), Gymnolaemata, Phylactolaemata, Amphineura, Gastropoda, Bivalvia, Oligochaeta, Hirudinea, Crustacea, Insecta, Chaetognatha, Ophiuroidesa, and Osteichthyes, and possibly other marine forms.
Sexual Reproduction — See Bisexual Reproduction.
Somatic Fertilization — Apparently only in Porifera.
Somatic Fission — Only in Protozoa (Sarcodina).
Sorites — Occur only in Porifera and possibly Anthozoa.
Spermatogenesis – Occurs in all male individuals that take part in Sexual Reproduction.
Spores — Occur only in Protozoa (but see also Agametes).
Sporogamy — Not clearly identified but presumably occurring only in Protozoa.
Sporogony — Occurs only in Protozoa (Sarcodina, Flagellata, and Sporozoa).
Sporozoites — Occur only in Protozoa (Sporozoa).
Sporulation — Same as Sporogony.
Statoblasts — Occur only in Phylactolaemata.
Staurogamy — Same as Cross-fertilization.
Stolonization — Occurs in some form in Hydrozoa, Scyphozoa, Anthozoa, Calyssozoa, Gymnolaemata, Phylactolaemata, Polychaeta, Crustacea, Pterobranchia, Ascidiacea, Thaliacea, and perhaps others.
    Gemmiparous Stolonization — Term used only in Polychaeta.

Strobilation — Occurs in Flagellata (reproductive bodies from adult), possibly in Ciliata, Scyphozoa (ephyrae or scyphistomae from strobilus), Anthozoa (polyps from polyp), Cestoda (proglottids from scolex), Polychaeta (adults or stolons from adult), and Ascidiacea (zooids from buds from zooid).

Successional Polyembryony — Occurs only in Trematoda.

Sycnhronogamy — Would occur in any hermaphrodite which does not show protandry or protogyny.

Syngamy — Occurs in most species of most groups, except for species using exclusively Hologamy, Conjugation, Nuclear Reorganization, Activation, Parthenogenesis, Asexual Reproduction, or any combination of these.

Syngony — Term used only in Nematoda.

Syntomy — Term used only in Polychaeta, except that it has been used for binary fission in the translated work of one author.

Syzygy — Occurs only in Protozoa (Sporozoa).

Thelytoky — Parthenogenesis producing only females. Occurs in Gastrotricha and Insecta.

Tomiparity — No examples of formal use of this word are known to us. (See Fission and Division)

Twinning — Same as Polyembryony (if identical).

Unisexual — With reference to reproduction, see Parthenogenesis. With reference to individuals, see Gonochorism.

Zoogamy — Sexual Reproduction. (See Bisexual Reproduction)

Zygogenesis — See Sexual Reproduction.

Appendix 1

GLOSSARY

Actinula — The larva developed from a planula in some Hydrozoa.

Activation — Initiation of development of an ovum; herein restricted to absence of fertilization (karyogamy).

Adult — The customary final form of the developing animal, usually connotes sexual maturity.

Agametes — Unicellular reproductive bodies that are diploid, do not require fusion, and are not gametes; animal spores.

Agamogenesis — Asexual reproduction.

Agamogony — Reproduction by means of agametes (not reproduction without gametes, which is asexual); also as sporogony.

Agamont — That form of an organism that produces agametes. (See also schizont)

Agamy I — Parthenogenesis.

Agamy II — Asexual reproduction (without gametes).

Agamy III — Reproduction by agametes.

Allogamy — Same as Outbreeding.

Ameiosis — Cell division without meiosis (reduction division); usually mitosis.

Ameiotic — Without meiosis; usually mitotic.

Ameiotic Parthenogenesis — Development of an ovum that has undergone no meiosis.

Amictic — Of females, producing ova incapable of being fertilized; of ova (eggs), developing parthenogenetically into females.

Amixis — Development without fertilization; same as apomixis.

Amoebae — Protozoans without fixed shape, with neither cilia nor flagella but moving by means of pseudopodia.

Amoebocyte — Motile cell in a metazoan that resembles an amoeba.

Amoeboid — Like an amoeba in form and movement.

Amoebulae — Tiny amoeboid cells, usually fragments that are agametes.

Amphiblastula — The free-swimming "larval" stage of some sponges in which the two poles differ in cell size and ciliation.

Amphigenesis — The mixing of gametes (genomes) from two individuals.

Amphigonous — Involving two sexes; cooperation of two individuals.

Amphigony — Reproduction involving two individuals in amphimixis.

Amphimictic — Reproducing by amphimixis.

Amphimixis — The mixing of two genomes by the union of two gametes.

Amphitoky — In Parthenogenesis, the production of both male and female forms.

Anabiosis — Any developmental process or condition in which the organism passes into a resting stage.

Ancestrula — In Gymnolaemata, the primary zooid that buds the colony.

Androgenesis I — The development of an egg having only a paternal nucleus.

Androgenesis II — Arrhenotoky.

Ansiogametes — Dissimilar gametes (spermatozoa and ova).

Anisogamous — Pertaining to anisogamy.

Anisogamy — Production of anisogametes — gametes of two types (ovum and spermatozoon).

Anthogenesis — Production of both males and females by parthenogenesis. (See also Deuterotoky)

Apogamete — "A gamete formed by apomixis." (See Ameiotic Parthenogenesis in the section on Parthenogenesis)
Apomictic — Involving apomixis; if sexual, involving meiosis but without karyogamy.
Apomixis — Reproduction without karyogamy.
Apozygote — Term proposed herein for an activated egg that develops without fertilization in parthenogenesis.
Architomy — Fragmentation with little or no advance formation of new organs. Includes frustulation, pedal laceration, schizometamery, and some other forms of epitoky.
Arrhenotoky — Parthenogenesis that produces only males.
Asexual — Not involving sex; specifically, not involving meiosis in reproduction; vegetative.
Asexual Reproduction — Process resulting directly in new (additional) individuals without meiosis.
Assembled Bodies — Reproductive bodies formed internally by aggregation of cells of several types. (See Gemmules, Sorites, Statoblasts, Podocysts, and Hibernacula)
Atokal — In epitoky, referring to the atoke, the portion of the worm which does not metamorphose (atokous).
Atoke — The unchanged (anterior and nonreproductive) portion of the worm in epitoky; sometimes also used for a nonreproductive individual.
Autocopulation — Self-insemination by direct transfer.
Autogamy — In Protozoa, the form of automixis involving fusion of sister nuclei which are still in the same cell.
Automictic — Involving automixis.
Automictic Parthenogenesis — Fusion of nongametic nucleus with the ovum nucleus.
Automixis — In Protozoa, activation by fusion of sister nuclei, either by autogamy or paedogamy.
Autotomy — Deliberate breaking off of an appendage or other structure; it is reproductive only if the fragment also regenerates into a new individual (dichotomous autotomy).
Autozooid — The feeding polyp of the bryozoan or other colony.
Avicularia — In Bryozoa, a structure or individual in the form of a bird's head or beak.
Axoblasts — The primary reproductive cells inside the axial cell of the dicyemid Mesozoa.

Basic Bisexual Reproduction — With reference to a complete reproductive cycle, it is bisexual reproduction unaltered in its genetic effects by any other reproduction in the cycle; thus, production of a single offspring by immediate biparental karyogamy.
Binary Fission — Fission of a cell into two, accompanied by nuclear division; in Protozoa or isolated cells of Metazoa.
Bisexual — Involving two sexes, either individuals of different sex or at least gonads of different sex. (See also Isogamy)
Bisexual Reproduction — Multiplication involving two individuals, or at least two gonads, producing the two types of gametes.
Blastostyle — The part of a hydrozoan gonangium which buds medusae.
Budding — A form of asexual reproduction, in which daughter individuals separate off after becoming independent; also the same process when the independent individuals do not separate. (See also Pallial Budding, Pyloric Budding, Secondary Budding, and Double Budding)
Buds — The structures or individuals formed by budding.

Calyx — The cup-like or vase-like body of a coelenterate polyp.
Cell Constancy — The production of all cells of an organ or an individual in the early embryo with no further cell divisions during later development, so that the body or organ consists of this fixed number of cells throughout its life. (See Eutely)
Cercaria — In Trematoda, a motile larval stage between the redia and the encysted metacercaria.
Chain of Gonophores — In strobilating animals, the chain of temporary structures carrying the gametes or the zygotes. (See Proglottids)
Ciliospore — A ciliated spore (as distinct from a flagellated one) produced by multiple fission.
Clonal — Produced by asexual processes from one parent; identical in genotype. (See Clone)
Clonal Fertilization — Fertilization by another member of the same clone.
Clone — The group of individuals formed by asexual processes from a single parent.
Coenurus Bladder — In Cestoda, a compound cysticercus consisting of a large bladder with internally budding scolices.
Colony — The interconnected group of individuals that were formed asexually from a single progenitor and its descendents.
Conjugation — In reproduction, temporary union of two protozoans; usually involving exchange of nuclear material. (See Cytogamy)
Copulation — The process of temporary physical union of two individuals to transfer spermatozoa.
Cormidium (pl. -dia) — In a siphonophore colony, a group of zooids attached to a stem.
Cross-fertilization — Union of gametes from two individuals, not from two gonads in one individual; usually assumed to be unrelated individuals.
Cycle — A sequence of reproductive processes; for the individual, extending from "birth" to "death"; for the species, including all the successive or alternative sequences that occur.
> Developmental Cycle — The life of one individual from the instant of origin to its death or end of its separate existence.
> Life Cycle — The same as sequence.
> Reproductive Cycle — The series of sequences involved in the production of an individual identical in form to the initial one.
Cyst — A sac or capsule containing a resting stage of a protozoan or a metazoan.
Cysticercus — In Cestoda, a larval stage consisting of a bladder with an invaginated developing scolex.
Cytogamy — In Protozoa, conjugation without exchange of nuclei.

Death — The cessation of life through failure of vital processes.
Deuterotoky — Parthenogenesis that produces both males and females.
Developmental Cycle — See Cycle.
Dichogamy — Hermaphroditism in which the male and female gametes mature at different times.
Dichotomous Autotomy — Division in which both parts do regenerate.
Dioecious — Having two sexes in the species but only one per individual.
Dioecism — In a species, the condition of having male and female individuals (or at least females); the opposite to monoecism.
Diploid — With a genotype consisting of two genomes; having twice the number of chromosomes found in a meiotically formed gamete.
Diploidy — The condition of being diploid.

Diporpa — A larva of certain trematodes formed by permanent fusion of two larvae.

Dissogeny — The condition of becoming sexually mature twice, both as a larva and as an adult, with degeneration of the larval gonads between the stages.

Dissogony — Same as Dissogeny.

Division — The breaking of a multicellular body into two or more. Includes fragmentation, polyembryony, strobilation, budding, epitoky, etc. (If a single cell, see Fission)

Egg — The female gamete (an unnecessary synonym of ovum, not used herein); the zygote formed by fusion of spermatozoon and ovum (or two isogametes), usually with special coverings; an embryo invested by an egg covering or shell; an apozygote, a parthenogenetic ovum from the instant of activation.

    Subitaneous — Summer egg, thin-shelled, parthenogenetic, and quick-hatching.

    Dormant — Winter eggs, diploid, thick-shelled, a resting egg.

    Amictic — Thin-walled eggs; eggs that cannot be fertilized.

    Mictic — Thin-walled; eggs that can be fertilized; some can also develop parthenogenetically.

    Composite — In certain trematodes, a zygote formed by fusion of two or more zygotes.

Embryo — The indefinite developmental stage from the start of cleavage either to attainment of a larval or juvenile stage capable of feeding itself, or to hatching, or to parturition.

Encystment — The processes of enclosing a cell, an individual protozoan, or a larva in a resistant covering or cyst.

Endogamy — Same as Inbreeding.

Endogenous — Formed on the inside.

Endomixis — A process of nuclear reorganization in which both macro- and micronuclei may be reformed from parts of the micronucleus.

Ephyrae — In Scyphozoa, the products of strobilation; each is a larval medusa.

Epigamy — Epitoky in which the entire preexisting atokous individual is modified, the posterior part to form the epitoke.

Episodes — The successive reproductive occurrences in an individual or pair.

Epitoke — The gonad-bearing fragment, or the sexual individual, formed by epitoky.

Epitoky — The processes in which all or part of the worm body changes in structure and physiology in preparation for sexual reproduction; may involve a detached gonophore or a fragmentation into two types of individuals. (See also Atoke and Epitoke)

Etheogenesis — Development from a spermatozoon alone (in plants only).

Eudoxy — The separation of a cormidium from certain siphonophore colonies to form a daughter colony.

Eutely — The absence of mitosis in the cells of an individual after the end of the cleavage phase, during which all necessary cells have been produced, resulting in a constant number of cells (or nuclei) in that individual throughout life. The term is also employed in animals where mitoses are continued in the gonads alone.

Exogamy — Same as Outbreeding.

Exogenous — Formed on the outside.

Female — An individual that produces ova or is of the sort that does produce them.

Fertilization — Fusion of a spermatozoon with an ovum, involving syngamy with karyogamy and activation.

    Clonal — Fusion of gametes from members of a clone.

    Fraternal — Fusion of gametes from sibling individuals.

Fission — The splitting of a cell into two (binary fission) accompanied by nuclear fission, or into many (multiple fission) after fission of the nucleus into many. (If multicellular, see Division)

Fissiparity — Reproduction by fission (unicellular) or division (multicellular). Same as Architomy and Paratomy together.

Foetus — Late stage in embryological development of vertebrates, usually mammals; also as fetus.

Fragmentation — General term for all division into two or more pieces without advance preparation at the site of the break; is reproductive only if both parts regenerate. (See Architomy and Paratomy)

Fraternal Fertilization — Fusion of gametes derived from sibling individuals.

Frustulation — Breaking off of an irregular piece from a polyp stalk, to form a nonciliated but otherwise planula-like larva.

Frustule — The piece broken off in frustulation.

Fusion — Structural union of two objects; in reproduction usually two gametes (syngamy) or two gamete nuclei (karyogamy).

Gametes — Unicellular reproductive bodies that are spermatozoa (always haploid), ova (usually haploid but sometimes diploid), isogametes (haploid), or hologametes (effectively haploid), formed by gametogenesis in Metazoa and also by multiple fission in Protozoa.

Gametic — Involving gametes.

Gametogenesis — The cytological production of gametes; includes both spermatogenesis and oogenesis.

Gamogenesis — Reproduction involving gametes.

Gamogony — Reproduction by means of gametes; usually used in Protozoa when there is multiple fission.

Gamont — An individual which produces gametes.

Gastrula — That stage in the early development of an embryo started when the archenteron is established.

Gemmation — Embryonic budding, in Insecta.

Gemmiparity — The production of a chain; same as strobilation.

Gemmiparous Stolonization — Producing a chain from a stolon.

Gemmulation — The production of gemmules (or other multicellular reproductive bodies).

Gemmule — Internal reproductive body consisting of cells of several types surrounded by a resistant covering; may develop, after release into either a larva or an adult.

Geneagenesis — Parthenogenesis.

Genome — One complete haploid set of chromosomes.

Genome Change — Alteration of the genome (gene complement) by addition, exchange, loss, or nuclear reorganization; in our context, represented by conjugation, hologamy, endomixis, autogamy, and hemimixis.

Genotype — One complete diploid set of chromosomes (two genomes), in a zygote or its derivative cells.

Germ Ball — Mass of cells, in some trematode embryos and larvae, produced by one of the blastomeres or its successors, each of which produces another embryo or larva containing its germ ball from which the next generation of larvae will be produced. (See Successional Polyembryony)

Germinal — Pertaining to a germ cell.

Gonad Transplant — Transfer of a primordial testis into a female, where it develops and produces spermatozoa to fertilize her ova. This results in an "artificial hermaphrodite", but the fertilizations are not self-fertilization because the gametes are from different genotypes.

Gonangium — A reproductive polyp in Hydrozoa.

Gonochorism — The condition of individuals which have the features of only one sex; cf. hermaphroditism.

Gonochoristic — Having only one sex per individual.

Gonochorist — An individual having only one sex.

Gonomery — Syngamy in which the chromosomes from the gametes do not mix but remain in separate groups for several divisions at least.

Gonophore — A detachable structure containing gonads or stored gametes or zygotes.

Gonozooid — Individual zooid whose chief function is reproduction.

Gymnospores — In gregarine Protozoa, products of multiple fission that then pair and fuse; similar to isogametes but derived from a fused pair of parents.

Gynandromorph — An individual, of a type not usually hermaphroditic, which is structurally part male and part female; usually male on one side and female on the other.

Gynandromorphism — The condition of having the body structurally part male and part female, without being hermaphroditic.

Gynogenesis I — Pseudogamy; development of ova without karyogamy.

Gynogenesis II — Production of females only, by parthenogenesis.

Haploid — Having a genotype consisting of one complete set of chromosomes — one genome; designated as N; cf. diploid.

Haploidy — The condition of having the haploid number of chromosomes.

Hectocotylus — Specialized tip of one of the arms of certain Cephalopoda, used to transfer spermatophores to the female and sometimes is broken off in the process.

Hemimixis — Nuclear reorganization in which the macronucleus breaks up and re–fuses.

Hemizygoid — Haploid.

Hemizygoid Parthenogenesis — Development from a haploid ovum, presumably with doubling of chromosomes at some point.

Hermaphrodite — An individual normally capable of producing both spermatozoa and ova, not necessarily simultaneously.

Hermaphroditic — Possessing reproductive organs of both sexes, either simultaneously or sequentially.

Hermaphroditism — The condition of an individual being an hermaphrodite; cf. Gonochorism.

Heterogamete — An anisogamete.

Heterogamy I — Same as Anisogamy.

Heterogamy II — In Insecta, the alternation or mixing of parthenogenesis with amphimixis.

Heterogenesis — Alternation of generations, usually sexual and asexual phases.

Heterogenotypic — Possessing in its various cells two or more genotypes that came from the fusion of blastomeres from two or more zygotes.

Heterogony — The alteration of parthenogenetic and zygogenetic generations.

Heterogyny — Polymorphism among the females of a species.

Heterospermic — In Merogony I, involving heterospermy.

Heterospermy — Activation of an enucleated egg by a spermatozoon from a different species.

Hibernaculum — Specially modified multicellular winter bud with thick "sclerotic" coat.

Hologamete — One of the entire individuals which fuse in hologamy.
Hologamous — Involving hologametes.
Hologamy — Reproduction of certain protozoans which fuse in the manner of gametes.
Homospermic — In Merogony I, involving homospermy.
Homospermy — Activation of an enucleated egg by a spermatozoon of the same species.
Hydatid Cyst — The coenurus larval stage of *Echinococcus* (Cestoda).
Hydranth — The feeding polyp in colonies of Hydrozoa.
Hydrocaulus — The branched stalk of hydrozoan colonies.
Hypodermic Insemination — Stabbing through the body wall of the female to inject spermatozoa.
Hypogenesis — Asexual reproduction.

Inbreeding — Amphimixis when the gametes are from closely related individuals (usually siblings).
Individual — An organized entity capable of performing integrated activities necessary to its continued existence and cycle.
Individuality — The condition of being an individual rather than part of an individual.
Infusorigen — A larva developed from a ball of cells inside a rhombogen, in Mesozoa.
Insemination — Introduction of semen (presumbly with spermatozoa) into the female.
Insemination by Spermatophore — Insemination with the spermatozoa packaged in a spermatophore.
Intersex — An abnormal individual, not hermaphroditic, which shows characters intermediate between those of the usual sexual individuals.
Isogametes — Gametes that are indistinguishable structurally, and therefore are not identifiable as spermatozoa and ova; they presumably have physiological sex distinctions.
Isogamous — Involving isogametes.
Isogamy — Bisexual reproduction by means of fusion of two isogametes.

Karyogamy — Fusion of gamete nuclei to produce a zygote nucleus (a synkaryon).

Larva — A developmental stage differing from an embryo in being able to secure its own nourishment; also differing substantially from the adult, which stage it attains by metamorphosis.
Life Cycle — The developmental sequence through which one organism passes; e.g., from fertilization to death, or from any other start to any other end. (See Tables 4 and 7 in Chapter 2)

Male — An individual that produces spermatozoa, or is of the sort that might produce them.
Medusae — The bell-shaped form of coelenterate individual; always a sexual form.
Medusoids — Individuals in siphonophore colonies supposed to be of the nature of medusae.
Meiosis — The processes that bring about segregation and reduction of the number of chromosomes to produce the haploid condition; reduction division(s).
Meiotic — Of the nature of reduction divisions.
Meiotic Parthenogenesis — Development of an ovum that is unfertilized but has undergone the normal meiotic divisions.
Merogametes — Gametes formed by multiple fission, in Protozoa.
Merogony I — Development of an enucleated egg fragment initiated by a spermatozoon.

Merogony II — Agametic production by merozoites produced by multiple fission, in Protozoa.

Merozoites — Nucleated cell fragments that are essentially spores without the protective cyst.

Metacercaria — In Trematoda, a cercaria larva which has encysted and metamorphosed into a juvenile fluke inside the cyst.

Metagenesis — Alternation of generations; usually sexual and asexual.

Metamorphosis — Change of form; usually the change from a larva into an adult.

Mictic — Involving mixing of genetic materials; capable of producing ova that can be fertilized; of a female, producing mictic ova; of an ovum, capable of being fertilized and then producing a female and if not fertilized, becoming a male; of an egg, derived from the mixing of gametes, eggs that were fertilized. (Used only in animals that also produce amictic eggs)

Miracidium — In Trematoda (Digenea) the free-swimming ciliated larva that hatches from the egg.

Mitotic — Involving mitosis; cell division in which the chromosome complement is unchanged.

Mixis — Mixing of genetic materials, usually by syngamy.

Monodisk — In strobilation, releasing the first new individual (or gonophore) before the second develops.

Monoecious — Having two sexes in the species but both in each individual.

Monoecism — The state of a species being monoecious; having two sexes in each individual.

Monogony — In certain species of Annelida, exclusively asexual reproduction, with no indication of sexuality in the species.

Monotomy — Simple binary fission, in Protozoa.

Multiple Fission — The splitting of a cell into many, after division of the nucleus into many.

Multiplication — Reproduction; production of additional individuals; cf. parareproduction.

Mutation — Heritable change in a genetic character resulting either from a change in one gene or alterations in chromosome structure.

Mychogamy — Self-fertilization.

Nematogen — That stage (in some Mesozoa) in which the axial cell produces agametes by endogenous division.
    Stem — The initial nematogen.
    Daughter or Ordinary — The ones developed from the agametes, which may later release infusorigens.

Neoteny — An evolutionary process in which reproduction comes to occur in the "larva" and the adult form is not usually attained. Compare Paedogenesis.

Neuters — Individuals lacking at least the major features of both sexes (nonreproductive individuals).

Nongametic — Not involving gametes.

Nuclear Reorganization — Any process that changes the nuclei of a protozoan, especially if meiosis is involved.

Nurse Cells — Cells from the ovary that accompany the ovum and nourish it. They may be sister oocytes of the ovum.

Onchomiracidium — In Trematoda (Monogenea), the larva that hatches from the egg.

Oncosphere — The first "larval" stage in Cestoda, a ciliated spheroidal sac.

Oocyst — A spore case formed by a zygote, in which numerous spores or sporozoites are formed.

Oogamy — The condition among anisogametes in which the ovum is much larger than the spermatozoon.

Oogenesis — The cytological production of ova.

Ookinete — A motile zygote in the parasitic life cycles of some Protozoa.

Outbreeding — Cross-breeding; fertilization by spermatozoa of an unrelated individual. In Basic Bisexual Reproduction, used to specify merely nonclonal fertilization.

Oviparous — Laying eggs.

Ovotestis — An organ that produces both spermatozoa and ova.

Ovoviviparous — Producing living young hatched from an egg retained inside the mother.

Ovum — Female gamete; usually large, subspherical, immobile, and yolky, haploid because of meiosis; may be fertilized or otherwise activated.

Paedogamous — Involving Paedogamy I, usually in autogamy.

Paedogamy — A form of automixis in Protozoa, in which the sister nuclei are now in daughter cells.

Paedogenesis — Sexual reproduction by nonadult individuals; larval reproduction, but can occur in any stage from the zygote to just before the adult. Compare Neoteny.

Palintomy — In Protozoa, repeated division producing spores; same as sporogony.

Pallial — Pertaining to the mantle.

Pansporoblast — A sporont containing two spores, in myxosporidian protozoans.

Parareproductive — Pertaining to processes involving sexuality and meiosis but not being multiplicative; i.e., conjugation, hologamy, nuclear reorganization.

Paratomy — Fragmentation in which the daughter individuals have grown into functioning units before separation. Includes budding, strobilation, and some epitoky.

Parthenogamy — Parthenogenetic development of a diploid cell.

Parthenogenesis — Development of an ovum without karyogamy.

    Diploid — The production of diploid individuals from an unfertilized but diploid ovum.

    Automictic — Development of a "zygote" formed by fusion of sister nuclei (autogamy or paedogamy).

    Haploid — The production of individuals from an unfertilized (haploid) ovum.

Parthenogenetic — Reproducing by parthenogenesis.

Patrogenesis — Development of an enucleated egg or fragment induced by fusion with a normal spermatozoon; Merogony I.

Pedal Laceration — Fragmentation of the foot of a polyp as it moves over a surface, followed by regeneration of the fragments into new polyps.

Planula — A free-swimming (ciliated) flattened and ovoid postblastula larva.

Plasmodium — A multinucleate mass initially formed either by fusion of cells and loss of plasma membranes or by multiple fission of a zygote nucleus without intervening plasma membranes.

Plasmogamy — Same as Plasmogony.

Plasmogony I — In Protozoa, the fusion of two cells (hologametes) without fusion of the nuclei.

Plasmogony II — Parthenogenesis in which the spermatozoon enters the egg but the nuclei remain separate.

Plasmotomy — Fission of a multinucleate protozoan into two or more parts without any nuclear division, the nuclei simply being distributed among the daughter cells. (Presumably the same as Multiple Fission)

Plastogamy — In Myxomycetes, the fusion of cells into a plasmodium.

Plerocercoid — A solid worm-like larval stage in some Cestoda.

Podocysts — Chitinous cysts in the pedal disk of certain scyphistoma larvae; they break away and develop into ciliated larvae.

Polar Body — One of the minute cells formed by mitosis during maturation of the oocyte.

Polar Body Fertilization — Fusion of a polar body nucleus with an ovum nucleus. (See Automictic Fertilization)

Polydisk — In strobilation, developing a stack or chain of individuals (or gonophores) before the first one is released.

Polyembryony — Production of two or more embryos by division of a zygote or embryo; may be by fragmentation or a process resembling budding.

    Successional — Production of a succession of generations of larvae by division of a germ ball and its progeny.

Polygenotypic — Describing an individual that consists of cells with unrelated genotypes, derived from two or more zygotes. (See Turbellaria)

Polymorphic — Appearing in more than one body form; polyp and medusa; male and female; egg, larva, and adult, etc.

Polyp — The cup- or vase-shaped form of coelenterate individuals.

Polyploidy — The condition in a cell of having more than the diploid usual number (2N) of chromosomes (three or more genomes)

Polypoids — Individuals in siphonophore colonies supposed to be of the nature of polyps.

Polyspermy — Entry of more than one spermatozoon into an ovum.

Prebuds — Stolon buds from which several generations of specialized buds are formed, in tunicates.

Prereproductive — Pertaining to processes that precede multiplication, such as reproductive behavior, insemination, and conjugation.

Progenesis — In Trematoda, parthenogenesis in the metacercaria larval stage.

Proglottids — The segments of the chain (strobilus) of a tapeworm.

Protandric — Hermaphroditic, reproducing by protandry.

Protandrous — Same as Protandric.

Protandry — In an hermaphrodite, the condition of producing spermatoza before the ova, or at least having the male gonads maturing before the female.

Protogamy — The fusion of gametes to form a binucleate zygote.

Protogenesis — Budding.

Protogynous — Hermaphroditic, reproducing by protogyny.

Protogyny — In an hermaphrodite, the condition of producing ova before the spermatozoa, or at least having the female gonads maturing before the male.

Pseudo-eggs — Cells given off from the surface of infusorigens (inside the rhombogen) to form ciliated infusoriform "larvae."

Pseudogamy — Parthenogenesis in which the spermatozoon enters and activates the egg without karyogamy.

Pseudo-planulae — A nonciliated planula-like larva.

Pseudovum — A female "gamete" formed without meiosis and therefore diploid.

Pyloric Budding — In Tunicata, the formation and fusion of two buds, one from the epicardial tubes and one from the oesophagus.

Rediae — In Trematoda, a larval stage between the sporocyst and the cercaria.

Reduction Bodies — In several groups, the regressed body of one individual in a resting stage. (See also Sorites, Gemmules, and Statoblasts.)

Regeneration — The replacement of cells, tissues, organs, or major body fragments which have been lost; reproductive only if two or more individuals result through regeneration of each fragment.

Reproduction — The origination of new organisms from preexisting ones; but specifically must be multiplicative, not merely changing of one or two existing individuals into new one(s).

Reproductive Bodies — Fragments (one cell or many) that separate to become, or participate in production of, new individuals.

Reproductive Cycle — See Cycle.

Resting Buds — Separated buds which develop only after a prolonged interval.

Rhombogen — A stage in the life history of some mesozoans derived either by transformation of nematogens or from agametes produced by nematogens.

Schizogamy I — In Annelida, epitoky in which only the posterior end is modified.

Schizogamy II — In Asteroidea, same as division.

Schizogenesis — Reproduction by splitting (fission or division).

Schizogony — In Sporozoa, the process of multiple fission when the nucleated fragments develop into adults, without passing through the spore or agamete stages.

Schizometamery — A form of epitoky, in which individual segments are fragmented out of the middle of a worm, regenerate both ends, multiply the segments, and then repeat, to produce up to 50 individuals.

Schizont — In Sporozoa, an organism that will give rise to others by splitting; usually a trophozoite.

Scissiparity — In multicellular animals, reproduction by division (either architomy or paratomy).

Scyphistoma — In Scyphozoa, the sessile hydroid larval stage which will strobilate.

Secondary Budding — The further budding from a newly-formed bud.

Self-fertilization — Union of spermatozoa and ova produced in gonads within an hermaphrodite to form a zygote. (See also Gonad Transplant)

Sequences — The cycles of reproductive processes, one or more cycle per species and each consisting of one or more process.

Sex — The phenomenon of certain animals existing in two forms that each contribute one set of hereditary factors to the next generation.

Sex Reversal — Change from the condition of being a functional male to that of female, or vice versa.

Sexual — Involving sex, in either one form or two.

Sexual Reproduction — Involving meiosis in gametogenesis.

Sexuality — The possession of sex.

Solenium — In Anthozoa, a gastrodermal tube that sometimes connects polyps in a colony.

Somatic — Involving the soma as distinct from the germ plasm.

Somatic Fertilization — Union of spermatozoa with cells in the female reproductive tract; not reproductive.

Somatic Fission — In Sarcodina without a visible nucleus, fission by simple constriction.

Sorites — In Porifera, assembled bodies similar to gemmules but without the protective covering.

Species — The kinds in which all animals exist; normally the potentially interbreeding individuals that are part of a single gene pool.

Sperm — See Spermatozoon.

Spermaries — In Coelenterata, the term for the simple testes that produce spermatozoa.

Spermatogenesis — The cytological production of spermatozoa.

Spermatophore — A sac-like structure or capsule used to transfer spermatozoa to the female.

Spermatozoa (sing. -zoon) — The mature male gametes.

Spore-cell — In Hydrozoa, an agamete that will give rise to a medusa. (See Sporogony II)

Spore-formation — See Sporulation.

Spores I — Unicellular reproductive bodies that are diploid and produced by multiple fission, generally with protective covering.

Spores II — In Hydrozoa, inappropriate term for frustules.

Sporocyst — A sac-like larval stage following the miracidium in the life history of many trematodes.

Sporogamy — In Sporozoa, said to be the formation of spores by an organism derived from a zygote. A contradictory term, as spores cannot be gametes and "all" spores "are" produced by organisms derived from a zygote.

Sporogony I — Reproduction by means of spores.

Sporogony II — In some Hydrozoa of either sex, undifferentiated germ cells in the epidermis become amoeboid and wander into the endoderm; they divide into two cells, one of which becomes enveloped by the other; the outer cell continues without division to provide protection and nourishment to the inner cell, the latter — a so-called spore-cell — divides and gives rise to a medusa.

Sporozoites — "Spores" without resistant covering, produced by sporogony.

Sporulation — Production of spores. (See Spores I)

Statoblasts — Internal masses of cells, each surrounded by a chitinous protective bivalve shell, that can survive the death of the parent colony and start a new one by budding.

Statocyst — Properly a balancing organ; the word has been cited as the same as statoblast.

Staurogamy — Cross-fertilization or outbreeding.

Stereogastrula — A solid gastrula, usually at the stage of a planula larva.

Stolonization — Production of a stolon; or production of a colony by budding from a stolon.

Stolon — A branch, either tubular or solid, connecting individuals in a colony, or from which buds develop; often said to be highly modified individuals.

Strobila — The chains formed by strobilation.

Strobilation — The formation of a chain by repeated division of the basal link, with the oldest link at the other end (polydisk); or formation of individuals in the same manner, each separating before the next one forms (monodisk).

Subitaneous — In animals that produce two types of eggs, refers to ones that are thin-shelled and destined to hatch rapidly; cf. Egg, thick-shelled.

Successional Polyembryony — Production of a succession of generations of larvae by successive fissions of its germinal daughter cells (germ ball) from the original zygote.

Swarmers — The numerous motile cells produced by multiple fission, they may be flagellated or ciliated or amoeboid.

Synchronogamy — Simultaneous production of ova and spermatozoa by an hermaphrodite.

Syncytium — Multinucleate protoplasmic mass in which cell membranes are not apparent.

Syngamy — The fusion of two gametes; usually implying fusion of the nuclei.

Syngony — In Nematoda, same as Hermaphroditism.

Synkaryon — The zygote nucleus resulting from fusion of the nuclei of two gametes.

Syntomy I — Multiple fission of Protozoa; the same as Schizogony.

Syntomy II — In Annelida, the same as Schizogamy I.
Syzygy — End to end attachment of two gregarine Protozoa, permanently but without cell fusion.

Thelytoky — Parthenogenesis in which only females are produced; Gynogenesis II.
Tomiparity — Reproduction by fission or division.
Trochophore — A larva shaped like a pear or spinning top and distinguished by an apical tuft of cilia and a trochus or ciliated equatorial band.
Trophozoites — The feeding stage of the sporozoan Protozoa, formed by metamorphosis of the sporozoites; usually same as schizont.
Twinning I (fraternal)—In Mammalia, simultaneous implantation of two or more zygotes and development of the progeny.
Twinning II (identical)—See Polyembryony.

Unisexual — Of an individual, possessing the structures of only one sex; of a species, consisting of individuals all of one sex (always female).
Unisexuality — In an individual, possession of the features of only one sex; in a species, the existence of females only.

Vermiform Bodies — Elongate buds in some suctorian Protozoa.
Viviparity — The production of embryos nourished (usually via a placenta) in the female reproductive tract and released as living young.
Viviparous — Producing young by viviparity.

Winter Buds — Reduction bodies in Bryozoa, or same as hibernacula.
Winter Eggs — Thick-shelled and designed to live over a winter before hatching.

Zoogamy — Sexual reproduction.
Zooids — The individual members of colonies as in Hydrozoa.
Zygogenesis — Reproduction by fusion of male and female gamete nuclei.
Zygogenetic — Reproducing by fusion of anisogametes.
Zygoid — Diploid.
Zygoid Parthenogensis — Development of an unfertilized ovum that is diploid (includes ameiotic parthenogenesis).
Zygoidy — The condition of being diploid (zygoid).
Zygote — The diploid cells resulting from fusion of gametes and therefore containing a synkaryon. (An activated but not fertilized "egg" would not be readily distinguishable but is herein termed an apozygote.)

Appendix 2

CLASSIFICATION AND ANNOTATION

The groups and species referred to in this monograph are listed here in a rough classification to help the reader to recognize them. If the information about each requires citation of authority or source, that is cited in the right-hand columns. Thus, this listing can be scanned for groups, for species, for authors cited, or for process reported. (The "groups" here cited are not taken from any formal classification but are used for recognition only; their status as phyla, classes, or other may be inferred, but this is not intended to be definitive.)

PROTOZOA

Sarcodina	Hyman, 1,119	(plasmotomy)
	Hyman, 1,130	(polyembryony)
	Vorontsova, Liosner, 5	(multiple budding)
Rhizopoda	Kudo, 5,496	(no sexual)
Proteomixida		
Nuclearia	Hyman, 1,127	(multiple fission)
Vampyrella	Hyman, 1,127	(multiple fission)
Amoebida (Lobosa)	Hyman, 1,125	(sporogony)
	Hyman, 1,119	(binary fission)
	Hyman, 1,130	(polyembryony)
	Vorontsova, Liosner, 5	(monotomy)
	Hyman, 1,130	(autogamy)
	Hyman, 1,130	(amoebid gametes)
	Hyman, 1,130	(anisogametes)
Protamoeba primitiva	Vorontsova, Liosner, 6	(somatic fission)
Sappinia	Hyman, 1,125	(plasmogony)
Sappinia diploidea	Hyman, 1,128	(encystment)
	Hyman, 1,125	(hologamy)
Endamoeba histolytica	Hyman, 1,128	(reproduction)
Paramoeba	Hyman, 1,127	(multiple fission)
Trichosphaerium	Hyman, 1,130	(agamogony)
	Hyman, 1,130	(isogametes)
Testacida		
Arcella	Hyman, 1,130	(multiple fission)
Foraminifera	Hyman, 1,133	(alternation)
Patellina	Hyman, 1,134	(hologamy)
Elphidium	Hyman, 1,133	(dimorphism)
Actinopoda		
Heliozoa		
Acanthocystis	Hyman, 1,138	(budding)
Actinophrys	Hyman, 1,138	(autogamy)
Actinosphaerium	Hyman, 1,138	(autogamy)
Radiolaria	Vorontsova, Liosner, 10	(colony duplic.)
Flagellata	Hyman, 1,89	(sporulation)
	Hyman, 1,89	(hologamy)
Phytomonadina	Hyman, 1,103	(isogamy)
Chrysomonadina	Vorontsova, Liosner, 11	(colony fragment.)
Volvox	Hyman, 1,105	(daughter colony)
Pandorina	Vorontsova, Liosner, 11	(colony reproduc.)
Eudorina	Vorontsova, Liosner, 11	(colony reproduc.)
Dinoflagellata	Hyman, 1,93	(encystment)
	Hyman, 1,98	(transv. fission)
Haplozoon	Hyman, 1,243	(strobilation)

Flagellata (*continued*)
    Dinoflagellata (*continued*)

	Noctiluca	Vorontsova, Liosner, 10	(multiple budding)
		Kudo, 5,376	(budding)
	Trypanosoma	Hyman, 1,109	(multiple fission)
	Trypanosoma lewisii	Vorontsova, Liosner, 11	(syntomy)
Opalinida (Protociliata)		Hyman, 1,183	(gametes)
		Kudo, 5,1028	(Basic Bisexual)
		Kudo, 5,1029	(longitud. fission)
		Hyman, 1,183	(plasmotomy)
		Vorontsova, Liosner, 6	(syntomy)
Ciliata		Hyman, 1,172	(asexual only)
		Kudo, 5,827	(no Basic Bisexual)
		Hyman, 1,172	(encystment)
		Hyman, 1,172	(transv. fission)
		Hyman, 1,172	(conjugation)
		Hyman, 1,176	(endomixis)
		Hyman, 1,177	(hemimixis)
		Hyman, 1,201	(longitud. fission)
		Kudo, 5,827	(asexual)
	Paramecium	Kudo, 5,242	(autogamy)
	Stylonychia	Vorontsova, Liosner, xviii, xx	(paratomy)
Astomata		Kudo, 5,955	(chains)
Chonotricha		Hyman, 1,202	(budding)
Peritricha		Hyman, 1,172	(longitud. fission)
	Vorticella	Vorontsova, Liosner, 6	(budding)
Suctoria		Hyman, 1,203	(no bisexual)
		Hyman, 1,203	(budding)
		Hyman, 1,203	(binary fission)
	Dendrosomides	Vorontsova, Liosner, 13	(branching)
	Lernaeophrya capitata	Mackinnon, Hawes, 250	(permanent conj.)
Sporozoa		Hyman, 1,144	(multiple fission)
		Hyman, 1,144	(agamogony)
		Hyman, 1,161	(apomixis only)
		Hyman, 1,161	(autogamy)
		Hyman, 1,161	(budding)
		Kudo, 5,627	(isogamy)
Telosporidia			
Gregarinida		Hyman, 1,145	(syzygy)
Haemosporidia			
Plasmodium			
Cnidosporidia		Hyman, 1,161	(autogamy)
Myxosporidia		Hyman, 1,161	(plasmotomy)

PORIFERA

(phylum in general)	Fell, 115	(fusion of larvae)
Calcarea	Hyman, 1,307	(reduction bodies)
	Hyman, 1,321	(reduction bodies)
	Hyman, 1,307	(colony duplic.)
	Hyman, 1,321	(anisogamy)
	Hyman, 1,321	(colony duplic.)
	Hyman, 1,321	(Basic Bisexual)
	Hyman, 1,321	(budding)
	Hyman, 1,321	(no gemmules)
Hexactinellida	Hyman, 1,331	(Basic Bisexual)
	Hyman, 1,331	(budding)
	Hyman, 1,331	(gemmules)
Demospongea	Hyman, 1,311	(Basic Bisexual)
	Hyman, 1,312	(self-fertiliz.)

	Hyman, 1,352	(gemmules)
	Hyman, 1,352	(reduction bodies)
Monaxonida		
Hadromerina		
Tethya maza	Vorontsova, Liosner, 15	(budding)
Haplosclerina		
Spongillidae	Hyman, 1,352	(reduction bodies)
	Hyman, 1,356	(gemmules)
Sclerospongea		
(class in general)	Storer, Usinger, 5,376	(reproduction unknown)

MESOZOA

(phylum in general)	Hyman, 1,233	(no Basic Bisexual)
Dicyemida	Hyman, 1,236+	(agamogony)
	Hyman, 1,236	(agamete fission)
	Hyman, 1,237	(pseudo-eggs)
Dicyemennea schulzianum	McConnaughey, 158	(parthenogenesis)
Orthonectida	Hyman, 1,243	(no Basic Bisexual)
	Hyman, 1,243	(hermaphroditism)
	Hyman, 1,243	(agamogony)

MONOBLASTOZOA

(phylum in general)	Hyman, 1,245	(no bisexual)
	Hyman, 1,245	(division)
Salinella salve	Hyman, 1,245	(only species)

PLACOZOA

(phylum in general)	Hyman, 1,243	(reprod. unknown)

COELENTERATA

(phylum in general)	Vorontsova, Liosner, 19	(asexual only)
Hydrozoa	Hyman, 1,409	(polyp buds)
	Hyman, 1,431	(longitud. divis.)
	Hyman, 1,423	(blastostyle buds)
	Hyman, 1,435	(polyembryony)
	Hyman, 1,431	(buds on medusa)
	Vorontsova, Liosner, 29	(pedal laceration)
	Hyman, 1,459	(gonophore buds)
	Hyman, 1,487	(colony duplicat.)
	Vorontsova, Liosner, 20	(transv. division)
Eleutheria	Hyman, 1,423	(self-fertilization)
Protohydra	Hyman, 1,440	(trans. & longit.)
Hydra	Hyman, 1,437	(transv. & longit.)
	Minchin, 148	(germinal budding)
Phialidium gastroblasta	Vorontsova, Liosner, 21	(longit. division)
Heterostephanus	Hyman, 1,403	(hydranth buds)
Hypolytus	Vorontsova, Liosner, 28	(fragmentation)
Moerisia	Vorontsova, Liosner, 26	(double buds)
Epenthesis macgradii	Vorontsova, Liosner, 27	(buds blastostyle)
Microhydra	Hyman, 1,459	(frustulation)
Schizocladium	Minchin, 146	("sporogony")
Margellium	Minchin, 147	(germinal budding)
Cunina	Minchin, 148	("sporogony")
Craspedacusta	Hyman, 1,459	(frustulation)
Trachylina	Hyman, 1,487	(buds larvae)
Anthomedusae	Hyman, 1,431	(buds from medusa)

Hydroza (*continued*)			
Narcomedusae		Vorontsova, Liosner, 28	(budding)
Siphonophora		Hyman, 1,402	(polymorphism)
		Kume, Dan, 98	(eudoxy)
Calycophora		Hyman, 1,487	(buds larvae)
		Hyman, 1,481	(buds colony)
Scyphozoa		Hyman, 1,522	(strobilation)
		Hyman, 1,507	(self-fertiliz.)
		Hyman, 1,528	(buds planulae)
		Hyman, 1,528	(podocysts)
		Hyman, 1,528	(fragmentation)
Semaeostomeae		Hyman, 1,528	(scyphistoma buds)
Pelagia		Hyman, 1,524	(no strobilation)
Chrysaora		Hyman, 1,507	(self-fertiliz.)
		Hyman, 1,529	(fragmentation)
		Hyman, 1,528	(podocysts)
		Hyman, 1,530	(pseudo-planulae)
Rhizostomeae			
Cassiopeia		Hyman, 1,528	(buds planulae)
Stauromedusae		Hyman, 1,529	(budding)
Haliclystus		Campbell, 181	(gastrula buds)
Lucernariidae		Vorontsova, Liosner, 33	(buds larvae)
Anthozoa		Hyman, 1,627	(self-fertiliz.)
Alcyonaria		Hyman, 1,592	(tentacle regen.)
Heliactis		Vorontsova, Liosner, 31	(fragmentation)
Stolonifera		Vorontsover, Liosner 32	(stolon buds)
Alcyonacea		Hyman, 1,543	(solenium buds)
Gorgonacea		Hyman, 1,543	(solenium buds)
Zoantharia			
Actiniaria		Vorontsova, Liosner, 30	(longit. division)
Actinia		Vorontsova, Liosner, 31	(pedal budding)
Gonactinia		Vorontsova, Liosner, 30	(polyp budding)
Gonactinia prolifera		Hyman, 1,590	(transv. division)
Boloceroides		Vorontsova, Liosner, 31	(tentacle bud)
Sagartia		Hyman, 1,590	(longit. division)
Metridium		Vorontsova, Liosner, 31	(fragmentation)
Madreporaria (Scleractinia)		Hyman, 1,590	(longit. division)
		Hyman, 1,607	(oral budding)
Fungia		Hyman, 1,609	(strobilation)

CTENOPHORA

(phylum in general)		Hyman, 1,692	(division)
		Hyman, 1,677	(dissogeny)
Tentaculata		Hyman, 1,677	(hermaphroditism)
		Giese, Pearse, 1,249	(self-fertiliz.)
Cydippida		Hyman, 1,679	(metamorphosis)
Platyctenea			
Gastrodes		Hyman, 1,689	(cydippid larva)
Ctenoplana		Hyman, 1,684	(males only)
		Hyman, 1,692	(fragmentation)
Coeloplana		Hyman, 1,692	(fragmentation)
NUDA		Hyman, 1,689	(development)
Beroe		Hyman, 1,677	(self-fertilization)

PLATYHELMINTHES

Nemertodermatida	Hyman, 2,129	(cross-fertiliz.)
Xenoturbellida	Ax, 215	(gamete source)
	Ax, 215	(fertilization external)

Gnathostomuloidea		Sterrer, 346	(reproduction)
		Sterrer, 346	(self-fertiliz.)
		Sterrer, 346	(fragmentation)
Turbellaria		Hyman, 2,125	(self-fertiliz.)
		Henley, 282	(parthenogenesis)
		Henley, 327	(gonomery)
		Hyman, 2,135	(egg polymorphism)
		Hyman, 2,180	(Basic Bisexual)
		Hyman, 2,234	(self-fertiliz.)
Acoela			
	Diopisthoporus	Hyman, 2,129	(hermaphroditism)
Rhabdocoela			
	Stenostomum	Vorontsova, Liosner, 36	(transv. division)
Catenulida		Hyman, 2,137	(chains)
Tricladida		Kume, Dan, 140	(composite eggs)
		Benazzi, 418	(pseudogamy)
	Phagocata velata	Hyman, 2,159	(fragmentation)
		Hyman, 2,180	(fragmentation)
	Dugesia	Vorontsova, Liosner, 36	(transv. division)
	Planaria	Vorontsova, Liosner, 36	(transv. division)
Temnocephaloidea		Hyman, 2,148	(common gonopore)
		Kaestner, 1,173	(self-fertilization)
Trematoda		Hyman, 2,239	(self-fertilization)
Monogenea		Smyth, 3,200	(self-fertilization)
	Polystoma integerrimum	Hyman, 2,244	(neoteny)
	Gyrodactylus	Hyman, 2,246	(polyembryony)
	Gyrodactylus elegans	Hyman, 2,246	(egg fusion)
	Sphyranura	Hyman, 2,243	(juvenile)
	Diplozoon	Hyman, 2,246	(diporpa larva)
Digenea		Smyth, 3,200	(self-fertilization)
		Hyman, 2,256	(polyembryony)
	Parorchis acanthis		
	Fasciola	Hegner, 220	(parthenogenesis)
	Fasciola hepatica		
Didymozoonidae		Hyman, 2,239	(dioecious)
Schistosomatidae		Hyman, 2,301	(dioecious)
		Hyman, 2,301	(polyembryony)
Cestoda		Vorontsova, Liosner, 35	(strobilation)
		Wardle, McLeod, 91	(cystic. divides)
		Vorontsova, Liosner, 38	(cystic. divides)
		Vorontsova, Liosner, 38	(cysticercus buds)
		Vorontsova, Liosner, 38	(coenurus buds)
	Urocystidium	Hyman, 2,342	(budding)
	Urocystis	Hyman, 2,342	(oncosphere buds)
Taenioidea		Hyman, 2,342	(cysticercus buds)
	Dioecocestus	Hyman, 2,326	(dioecious)
	Echinococcus	Smyth, 5	(polyembryony)
		Vorontsova, Liosner, 38	(hydatid budding)
	Multiceps	Smyth, 5	(polyembryony)
	Taenia crassiceps	Hyman, 2,342	(cysticercus buds)
Pseudophyllidea			
	Sparganum prolifer	Hyman, 2,341	(plerocercoid bud)
	Pseudophyllidae	Hyman, 2,340	(fragmentation)
CESTODARIA		Hyman, 2,326	(hermaphroditism)

RHYNCHOCOELA

(phylum in general)		Gibson, 80	(ovotestis)
	Lineus	Hyman, 2,516	(fragmentation)

ACANTHOCEPHALA

(phylum in general)	Hyman, 3,3	(dioecious)

ROTIFERA

Seisonidea	Hyman, 3,107	(dioecious)
	Hyman, 3,107	(one egg form)
Bdelloidea	Hyman, 3,110	(all females)
	Hyman, 3,137	(no asexual)
Monogononta	Hyman, 3,100	(Basic Bisexual)
	Hyman, 3,100	(parthenogenesis)
	Hyman, 3,100	(egg polymorphism)
	Hyman, 3,137	(no asexual)

GASTROTRICHA

Macrodasyoidea		
Dactylopodalia	Hyman, 3,164	(sex arrangement)
Chaetonotoidea	Hyman, 3,167	(parthenogenesis)
	Hyman, 3,167	(no asexual)
Neodasys	Hyman, 3,167	(possibly bisexual)
Xenotrichula	Hyman, 3,167	(possibly bisexual)

KINORHYNCHA

(phylum in general)	Hyman, 3,180	(dioecious)

PRIAPULOIDEA

(phylum in general)	Hyman, 3,191	(dioecious)
	Hyman, 3,195	(gametes free)

NEMATODA

(phylum in general)	Hyman, 3,243	(mostly dioecious)
	Hyman, 3,245	(self-fertiliz.)
Mermis	Hyman, 3,245	(parthenogenesis)
Heterodera	Hyman, 3,245	(parthenogenesis)
Rhabditis	Hyman, 3,245	(parthenogenesis)
	Hyman, 3,245	(activation)
Mesorhabditis belari	Giese, Pearse, 1,397	(pseudogamy)
Trichinella spiralis		

GORDIOIDEA (Nematomorpha)

Gordioidea	Hyman, 3,461	(dioecious)
Nectonematoidea	Hyman, 3,461	(dioecious)

CALYSSOZOA (Endoprocta)

(phylum in general)	Hyman, 3,534	(hermaphroditism)
	Hyman, 3,536	(budding)
Loxosomatidae	Hyman, 3,545	(embryonic buds)
Loxosoma	Vorontsova, Liosner, 51	(buds released)

BRYOZOA

Phylactolaemata	Hyman, 5,451	(self-fertiliz.)
	Hyman, 5,453	(clonal fertiliz.)
	Hyman, 5,455	(budding)

	Vorontsova, Liosner, 48	(statoblasts)
	Hyman, 5,408	(hibernacula)
	Patters, 399	(polyembryony)
Gymnolaemata	Hyman, 5,345	(sperm life)
	Hyman, 5,345	(self-fertiliz.)
	Hyman, 5,357	(ancestrula buds)
	Hyman, 5,362	(stolon buds)
	Patterson, 399	(embryo buds)
	Vorontsova, Liosner, 45	(embryo buds)
Cyclostomata	Vorontsova, Liosner, 45	(polyembryony)
Crisia	Hyman, 5,345	(parthenogenesis)
Cheilostomata		
Bugula	Hyman, 5,345	(self-fertiliz.)
Membranipora	Hyman, 5,354	(division)

PHORONIDA

(phylum in general)	Hyman, 5,249	(hermaphroditism)
	Hyman, 5,262	(regeneration)
	Kume, Dan, 242	(coelomic fertil.)
Phoronis muelleri	Hyman, 5,251	(coelomic fertil.)
Phoronis ovalis	Hyman, 5,262	(division)

BRACHIOPODA

Inarticulata	Hyman, 5,562	(dioecious)
Lingula	Hyman, 5,570	(no metamorphosis)
Articulata	Hyman, 5,522	(mostly dioecious)
Argyrotheca	Hyman, 5,562	(hermaphroditism)
Pumilus	Williams, Rowell, 43	(hermaphroditism)
Platidia	Williams, Rowell, 43	(hermaphroditism)

MOLLUSCA

Monoplacophora	Hyman, 6,151	(dioecious)
Amphineura (Polyplacophora)	Hyman, 6,111	(mostly dioecious)
Trachydermon raymondi	Hyman, 6,111	(hermaphroditism)
	Hyman, 6,111	(self-fertiliz.)
Solenogastres (Aplacophora)	Hyman, 6,53	(mostly hermaphr.)
Chaetodermatidae		
Chaetoderma	Barnes, 1,250	(dioecious)
Gastropoda	Hyman, 6,297	(self-fertiliz.)
	Hyman, 6,492	(self-fertiliz.)
	Wilson, 433	(gonomery)
Prosobranchia	Hyman, 6,297	(parthenogenesis)
	Hyman, 6,599	(protandrous)
Crepidula fornicata	Morton, 133	(sex changes)
Opisthobranchia	Hyman, 6,599	(self-fertiliz.)
	Hyman, 6,599	(hermaphroditism)
Pulmonata	Hyman, 6,599	(self-fertiliz.)
	Hyman, 6,599	(hermaphroditism)
Lymnaea	Morton, 135	(self-fertiliz.)
Bivalvia	Barnes, 3,406	(mostly dioecious)
Pecten	Barnes, 2,344	(hermaphroditism)
	Barnes, 3,406	(ovotestis)
Ostrea	Borradaile, Potts, 630	(hermaphroditism)
Anodonta	Borradaile, Potts, 630	(hermaphroditism)
Sphaeriidae	Barnes, 3,406	(self-fertiliza.)
Scaphopoda	Barnes, 3,410	(dioecious)
Cephalopoda	Barnes, 3,427	(hermaphroditism)

SIPUNCULOIDEA

(phylum in general)	Hyman, 5,652	(mostly dioecious)
	Hyman, 5,653	(self-fertiliz.)
	Hyman, 5,661	(no asexual)
	Hyman, 5,652	(parthenogenesis)

ECHIUROIDEA

Echiurida	Barnes, 3,683	(dioecious)
Bonellia	Barnes, 3,683	(parasitic male)
Saccosomatida		

MYZOSTOMIDA

(phylum in general)	Kaestner, 1,510	(Basic Bisexual)
Myzostomum	Kaestner, 1,510	(spermatophore)

ANNELIDA

Polychaeta	Barnes, 3,274	(mostly dioecious)
	Barnes, 3,275	(division)
	Kaestner, 1,474	(multiple epitoky)
	Pennak, 461	(schizogamy)
Syllidae	Kaestner, 1,474	(epitoky)
	Adams, MS	(strobilation)
Syllis	Barnes, 3,276	(budding)
	Dales, 164	(epitoky)
	Vorontsova, Liosner, 44	(budding)
Syllis ramosus	Barnes, 3,276	(epitoky)
Syllis hyalina	Dales, 164	(epitoky)
Syllis gracilis	Vorontsova, Liosner, 40	(epitoky)
Autolytus	Kaestner 1,476	(strobilation)
Autolytus prolifer	Dales, 159	(epitoky)
Trypanosillis	Barnes, 3,276	(budding)
Exogone gemmipara	Vorontsova, Liosner, 43	(epitoky)
Sabellidae (fanworms)	Barnes, 3,277	(hermaphroditism)
	Barnes, 3,274	(asexual)
	Barnes, 3,277	(self-fertiliz.)
Sabella	Dales, 158	(epitoky)
Chaetopteridae		
Chaetopterus	Dales, 152	(epitoky)
Phyllochaetopterus	Barnes, 3,274	(fragmentation)
Cirratulidae		
Zeppelina monostyla	Kaestner, 1,475	(asexual only)
Dodecaceria caulleryi	Schroeder, Hermans, 7	(schizometamery)
	Vorontsova, Liosner, 41	(epitoky)
Nereidae		
Nereis irrorata	Barnes, 3,275	(fragmentation)
Platynereis spp.	Schroeder, Hermans, 3,48	(epitoky)
Eunicidae		
Eunice schemacephala	Barnes, 3,277	(epitoky)
Palolo siciliensis	Kaestner, 1,474	(epitoky)
Tylorrhynchus	Dales, 165	(epitoky)
Nephtyidae		
Nephtys	Kaestner, 1,473	(epitoky)
Oligochaeta	Barnes, 3,294	(all hermaphrod.)
	Barnes, 3,299	(chains)
	Stephenson, 478	(budding)
	Stephenson, 542	(polyembryony)

Lumbriculidae		
Lumbriculus variegatus	Vorontsova, Liosner, 39	(asexual only)
Tubificidae		
Limnodrilus udekemianus	Barnes, 3,298	(self-fertiliz.)
Naididae	Barnes, 3,299	(asexual only)
Nais	Barnes, 3,299	(division)
Aulophorus	Barnes, 3,299	(division)
Allonais paraguayensis	Barnes, 3,299	(fragmentation)
Lumbricidae		
Lumbricus trapezoides	Vorontsova, Liosner, 39	(gastrula division)
Allolobophora caliginosa	Stephenson, 542	(embryo division)
Aeolosomatidae	Barnes, 3,294	(no dist. gonads)
Hirudinea	Barnes, 3,311	(hermaphroditism)
	Mann, 104	(hermaphroditism)
	Barnes, 3,311	(self-fertiliz.)
	Barnes, 3,311	(no asexual)
Rhynchobdellida		
Gnathobdellida		
Pharyngobdellida		
Herpobdella		
Acanthobdellida		
Archiannelida		
Protodrilus	Borradaile, Potts, 313	(hermaphroditism)

DINOPHILOIDEA

(phylum as a whole)	Kaestner, 1,507	(fraternal fert.)

TARDIGRADA

Heterotardigrada	Barnes, 3,689	(dioecious)
	Barnes, 3,689	(parthenogenesis)
	Barnes, 3,689	(eutely)
	Pennak, 249	(anabiosis)
Echiniscus	Barnes, 2,582	(no males)
Eutardigrada	Pennak, 247	(Basic Bisexual)
	Pennak, 249	(no asexual)

PENTASTOMIDA

(phylum in general)	Cheng, 551	(dioecious)
	Barnes, 3,691	(dioecious)

ONYCHOPHORA

(phylum in general)	Storer, Usinger, 4,524	(larvae)
	Parker, Haswell, 1,530	(viviparity)
	Barnes, 2,568	(fertilization)

ARTHROPODA

Merostomata	Barnes, 3,456	(Basic Bisexual)
Pycnogonida	Barnes, 3,508	(Basic Bisexual)
Arachnida	Lapage, 164	(parthenogenesis)
	Barnes, 3,464	(dioecious)
Phalangium opilio	Cloudsley-Thompson, 144	(parthenogenesis)
Crustacea		
Branchiopoda	Barnes, 2,445	(parthenogenesis)
Cladocera	Barnes, 2,445	(alternation)
	Waterman, 12	(paedogenesis)

Crustacea (*continued*)
    Isopoda   Borradaile, Potts, 405   (hermaphroditism)
    Ostracoda   Barnes, 2,445   (parthenogenesis)
    Copepoda   Barnes, 3,520   (hermaphroditism)
         Wilson, 433   (gonomery)
    Branchiura
    Cirripedia   Barnes, 3,520   (hermaphroditism)
         Waterman, 1,12   (paedogenesis)
        Scalpellum   Waterman, 1,412   (parthenogenesis)
        Ibla   Waterman, 1,412   (parthenogenesis)
        Trypetesa   Waterman, 1,412   (parthenogenesis)
    Rhizocephala   Waterman, 1,12   (budding)
         Waterman, 1,12   (stolonization)
         Waterman, 1,412   (self-fertiliz.)
        Peltogasterella   Barnes, 2,467   (gonad transplant)
    Malacostraca   Cameron, 170   (hermaphroditism)
Pauropoda   Kaestner, 2,432   (dioecious)
Symphyla   Barnes, 3,657   (Basic Bisexual)
     Barnes, 3,657   (parthenogenesis)
Diplopoda   Barnes, 3,662   (Basic Bisexual)
Chilopoda   Barnes, 3,655   (Basic Bisexual)
Insecta   Hagan, 24   (polyspermy)
     Davey 62   (parthenogenesis)
     Patterson, 399   (polyembryony)
    Orthoptera
        Acrididae   Imms, 8,192   (polyembryony)
    Plecoptera   Imms, 5,166   (hermaphroditism)
    Embioptera   Essig, 178   (parthenogenesis)
    Homoptera
        Coccidae
            Icerya purchasi   Hagen, 14   (polyspermy)
        Aphididae   Hagen, 15   (heterogamy)
    Anoplura
        Pediculus   Imms, 2,144   (hermaphroditism)
    Diptera
        Miastor   Essig, 730   (paedogenesis)
        Chironomus grimmii   Hagen, 56   (paedogenesis)
    Cynipidae   Hagen, 15   (heterogamy)
    Coleoptera
        Micromalthus   Blackwelder, MS   (paedogenesis)
        Graphognathus leucoloma   Blackwelder, MS   (oblig. parthenog.)
    Strepsiptera   Davey, 62   (polyembryony)
    Hymenoptera   Patterson, 399   (polyembryony)

CHAETOGNATHA

(phylum in general)   Hyman, 5,30   (hermaphroditism)
     Hyman, 5,44   (regeneration)
     Hyman, 5,30   (spermatophore)
    Sagitta   Hyman, 5,29   (self-fertiliz.)

POGONOPHORA

(phylum in general)   Ivanov, 90   (spermatophore)
     Hyman, 5,219   (dioecious)
     Ivanov, 90   (dioecious)

ECHINODERMATA

(phylum in general)   Patterson, 399   (polyembryony)
Crinoidea   Hyman, 4,71   (dioecious)

		Hyman, 4,75	(marsupium)
		Swan, 421	(regeneration)
		Fell, 51	(regeneration)
	Isometra vivipara	Hyman, 4,75	(internal fertil.)
Somasteroidea		Hyman, 4,317	(reprod. unknown)
	Platasterias	Spencer, Wright, U41	(reprod. unknown)
Asteroidea		Hyman, 4,288	(hermaphroditism)
		Hyman, 4,288	(sex reversal)
		Hyman, 4,288	(self-fertiliz.)
		Hyman, 4,310	(division)
		Swan, 418	(schizogamy)
	Linckia	Hyman, 4,313	(autotomy)
Ophiuroidea		Hyman, 4,623	(dioecious)
		Hyman, 4,623	(sex reversal)
		Hyman, 4,625	(brooding)
		Hyman, 4,625	(parthenogenesis)
		Hyman, 4,623	(self-fertiliz.)
		Hyman, 4,625	(hermaphroditism)
	Amphipholis squamata	Hyman, 4,623	(hermaphroditism)
	Ophiactis	Hyman, 4,642	(autotomy)
Echinoidea		Hyman, 4,478	(ovotestis)
		Hyman, 4,479	(self-fertiliz.)
		Harvey, 198	(parthenogenesis)
Holothurioidea		Hyman, 4,171	(mostly dioecious)
		Hyman, 4,202	(fragmentation)
		Hyman, 4,176	(self-fertiliz.)
		Vorontsova, Liosner, 53	(multiple fragm.)
		Easton, 587	(budding)
Dendrochirota		Hyman, 4,171	(hermaphroditism)
	Cucumaria	Hyman, 4,171	(hermaphroditism)
Synaptida		Hyman, 4,171	(hermaphroditism)

PTEROBRANCHIA

Cephalodiscus	Hyman, 5,175	(stalk buds)
Cephalodiscus sibogae	Hyman, 5,171	(males & neuters)
Atubaria	Hyman, 5,179	(no males)
Atubaria heterolopha	Hyman, 5,179	(noncolonial)
Rhabdopleura	Hyman, 5,181	(stolon buds)

ENTEROPNEUSTA

(phylum in general)	Hyman, 5,121	(Basic Bisexual)
Balanoglossus capensis	Hyman, 5,132	(fragmentation)

PLANCTOSPHAEROIDEA

(phylum in general)	Hyman, 5,193	(adults unknown)
	Hyman, 5,193	(reprod. unknown)

TUNICATA

Larvacea		Borradaile, Potts, 732	(mostly hermaphr.)
		Vorontsova, Liosner, 54	(no asexual)
	Oikopleura dioica	Parker, Haswell, 7,46	(dioecious)
Ascidiacea		Borradaile, Potts, 4,726	(not self-fertile)
		Storer, Usinger, 535	(mostly dioecious)
		Barnes, 3,815	(budding)
		Vorontsova, Liosner, 61	(budding)
		Vorontsova, Liosner, 57	(colony duplic.)
		Vorontsova, Liosner, 65	(resting buds)

Ascidiacea (*continued*)
    Distaplia — Vorontsova, Liosner, 58 — (fragmentation)
    Trididemnum — Vorontsova, Liosner, 57,61 — (pyloric budding)
        Vorontsova, Liosner, 60 — (dual budding)
  Diazonidae — Kume, Dan, 570 — (strobilation)
    Diazona — Vorontsova, Liosner, 65 — (fragmentation)
  Polyclinidae — Barnes, 3,816 — (fragmentation)
    Vorontsova, Liosner, 64 — (fragmentation)

Thaliacea — Parker, Haswell, 7,37 — (Basic Bisexual)
  Parker, Haswell, 7,37 — (hermaphroditism)
  Parker, Haswell, 7,39 — (budding)
  Pyrosomata — Vorontsova, Liosner, 66 — (fragmentation)
    Pyrosoma
  Doliolida
    Doliolum
  Salpida — Parker, Haswell, 7,43 — (clonal fertiliz.)
    Salpa

CEPHALOCHORDATA

(phylum in general) — Parker, Haswell, 7,59 — (mostly dioecious)
  Breder, Rosen, 13 — (hermaphrod. rare)

VERTEBRATA

Agnatha — Parker, Haswell, 7,194 — (fertilization)
  Parker, Haswell, 7,194 — (Basic Bisexual)
  Myxine glutinosa — Breder, Rosen, 13 — (self-fertiliz.)
Chondrichthyes — Breder, Rosen, 28 — (Basic Bisexual)
  Parker, Haswell, 7,222 — (dioecious)
  Gresson, 49 — (hermaphroditism)
  Raja batis — Wilson, 433 — (gonomery)
Osteichthyes — Breder, Rosen, 655 — (Basic Bisexual)
  Darevsky, 119 — (parthenogenesis)
  Fundulus — Wilson, 433 — (gonomery)
  Symbranchidae — Breder, Rosen, 655 — (protogyny)
  Sparidae — Breder, Rosen, 661 — (hermaphroditism)
  Maenidae — Breder, Rosen, 661 — (hermaphroditism)
  Serranidae — Breder, Rosen, 401 — (self-fertiliz.)
  Breder, Rosen, 657 — (self-fertiliz.)
Amphibia — Parker, Haswell, 7,452 — (spermatophore)
  Darevsky, 119 — (parthenogenesis)
  Parker, Haswell, 7,452 — (Basic Bisexual)
  Gresson, 49 — (hermaphroditism)
  Patterson, 399 — (polyembryony)
  Mexican axolotl — Parker, Haswell, 7,456 — (paedogenesis)
  Cryptobranchus — Wilson, 433 — (gonomery)
Reptilia — Parker, Haswell, 7,483 — (Basic Bisexual)
  Patterson, 399 — (polyembryony)
  Chelonia — van Tienhoven, 69 — (hermaphroditism)
    Lacerta saccicola — Darevsky, 115 — (parthenogenesis)
    Cnemidophorus — Darevsky, 119 — (parthenogenesis)
Aves — Parker, Haswell, 6,598 — (Basic Bisexual)
  Darevsky, 119 — (parthenogenesis)
  Patterson, 398 — (polyembryony)
  Gresson, 49 — (hermaphroditism)
  Domestic fowl — Darevsky, 128 — (parthenogenesis)
Mammalia — Darevsky, 119 — (parthenogenesis)
  Gresson, 49 — (hermaphroditism)
  Armadillos — Patterson, 399 — (polyembryony)
  Humans — Patterson, 399 — (polyembryony)

REFERENCES

The following are the works actually cited in the text as sources of specific information about reproductive processes. Usually only a single reference is given even though the data were found in several. Many other works were examined, either for verification or for details of life cycle and manner of living.

Adams, J. K. V., unpublished manuscript.
Ax, P., Relationships and phylogeny of the Turbellaria, in *The Lower Metazoa,* Dougherty, E. C., Brown, Z. N., Hanson, E. D., and Hartman, W. D., Eds., University of California Press, Berkeley, 1963, chap. 14.
Barnes, R.D., *Invertebrate Zoology,* 1st, 2nd, and 3rd eds., W. B. Saunders, Philadelphia, 1963, 1968, and 1974.
Benazzi, M., Genetics of reproductive mechanisms and chromosome behavior in some fresh-water triclads, in *The Lower Metazoa,* Dougherty, E. C., Brown, Z. N., Hanson, E. D., and Hartman, W. D., Eds., University of California Press, Berkeley, 1963, chap. 30.
Blackwelder, R. E., personal communication.
Borradaile, L. A. and Potts, F. A., *The Invertebrata,* 4th ed., Kerkut, G. A., Ed., Cambridge University Press, London, 1961.
Breder, C. M., Jr. and Rosen, D. E., *Modes of Reproduction in Fishes,* Natural History Press, Garden City, N.Y., 1966.
Cameron, T. W. M., *Parasites and Parasitism,* Methuen, London, 1956.
Campbell, R. D., Cnidaria, in *Reproduction of Marine Invertebrates,* Vol. 1, Giese, A. C. and Pearse, J. S., Eds., Academic Press, New York, 1974, chap. 3.
Cheng, T. C., *The Biology of Animal Parasites,* W. B. Saunders, Philadelphia, 1964.
Cloudsley-Thompson, J. L., *Spiders, Scorpions, Centipedes and Mites,* Pergamon Press, London, 1958.
Dales, R. P., *Annelids,* 2nd ed., Hutchinson University Library, London, 1967.
Darevsky, I. S., Natural parthenogenesis in a polymorphic group of caucasian rock lizards related to *Lacerta saxicola,* Eversmann, *J. Ohio Herpetol. Soc.,* 5, 115, 1966.
Davey, K. G., *Reproduction in the Insects,* Oliver and Boyd, Edinburgh, 1965.
Easton, W. H., *Invertebrate Paleontology,* Harper & Bros., New York, 1960.
Essig, E. O., *College Entomology,* Macmillan, New York, 1942.
Fell, H. B., Ecology of crinoids, in *Physiology of Echinodermata,* Boolootian, R. A. Ed., Interscience Press, New York, 1966, chap. 2
Fell, P. E., Porifera, in *Reproduction of Marine Invertebrates,* Vol. 1, Giese, A. C. and Pearse, J. S., Eds., Academic Press, New York, 1974, chap. 2.
Gibson, R., *Nemerteans,* Hutchinson University Library, London, 1972.
Giese, A. C. and Pearse, J. S., *Reproduction of Marine Invertebrates,* Vol. 1, (1974), Vol. 2, Vol. 3, Academic Press, New York, 1975.
Gray, P., *Dictionary of the Biological Sciences,* Reinhold Publishing, New York, 1967.
Gresson, R. A. R., *Essentials of General Cytology,* University of Edinburgh Press, Edinburgh, 1948.
Hagan, H. R., *Embryology of the Viviparous Insects,* Ronald Press, New York, 1951.
Harvey, E. B., *The American Arbacia and Other Sea Urchins,* Princeton University Press, Princeton, 1956.
Hegner, R. W., *Invertebrate Zoology,* Macmillan, New York, 1933.
Henley, C., Platyhelminthes (Turbellaria), in *Reproduction of Marine Invertebrates,* Vol. 1, Giese, A. C. and Pearse, J. S., Eds., Academic Press, New York, 1974, chap. 5.
Hyman, L. H., *The Invertebrates,* Vol. 1 (1940), Vol. 2 (1951), Vol. 3 (1951), Vol. 4 (1955), Vol. 5 (1959), Vol. 6 (1967), McGraw-Hill, New York.
Imms, A. D., *A General Textbook of Entomology,* 5th ed., Methuen and Co., London, 1942.
Imms, A. D., *A General Textbook of Entomology,* 2nd ed., E. P. Dutton, New York, 1930; 8th ed., Methuen, London, 1951.
Ivanov, A. V., *Pogonophora,* Academic Press, London, 1963.
Kaestner, A., *Invertebrate Zoology,* Vol. 1 (1967), Vol. 2 (1968), Vol. 3 (1970), Interscience, New York.
Kudo, R. R., *Protozoology,* 5th ed., Charles C Thomas, Springfield, 1966.
Kume, M. and Dan, K., *Invertebrate Embryology,* NOLIT, Belgrade, 1968.
Lapage, G., *Parasitic Animals,* Cambridge University Press, London, 1951.
Mackinnon, D. L. and Hawes, R. S. J., *An Introduction to the Study of Protozoa,* Oxford University Press, London, 1961.
Mann, K. H., *Leeches (Hirudinea),* Pergamon Press, London, 1961.

McConnaughey, B. H., Mesozoa, in *The Lower Metazoa,* Dougherty, E. C., Brown, Z. N., Hanson, E. D., and Hartman, W. D., Eds., University of California Press, Berkeley, 1963, chap. 11.
Meglitsch, P. A., *Invertebrate Zoology,* Oxford University Press, New York, 1972.
Minchin, E. A., Hydromedusae, in *Encyclopedia Britannica,* 12th ed., Vol. 14, 1910, 135.
Morton, J. E., *Molluscs,* 4th ed., Hutchinson, London, 1967.
Parker, T. J. and Haswell, W. A., *A Text-book of Zoology,* Vol. 1, 6th ed., Lowenstein, O., Macmillan, London, 1965; Vol. 2, 7th ed., Marshall, A. J., Macmillan, London, 1962.
Patterson, J. T., Polyembryony in animals, *Q. Rev. Biol.,* 2, 399, 1927.
Pennak, R. W., *Fresh-water Invertebrates of the United States,* Ronald Press, New York, 1953.
Schroeder, P. C. and Hermans, C. O., Annelida: Polychaeta, in *Reproduction of Marine Invertebrates,* Vol. 3, Giese, A. C. and Pearse, J. S., Eds., Academic Press, New York, 1974, chap. 1.
Sadleir, R. M. F. S., *The Reproduction of Vertebrates,* Academic Press, New York, 1973.
Smyth, J. D., *The Physiology of Trematodes,* W. H. Freeman, San Francisco, 1966.
Sonneborn, T. M., Breeding systems, reproductive methods and species problems in Protozoa, in *The Species Problem,* Publication No. 50, Mayr, E., Ed., American Association for the Advancement of Science, Washington, D.C. 1957, 155.
Spencer, W. K. and Wright, C. W., *Treatise of Invertebrate Paleontology,* (Part U), University of Kansas Press, Lawrence, 1966.
Stephenson, J., *The Oligochaeta,* Oxford University Press, Oxford, 1930.
Sterrer, W., Gnathostomulida, in *Reproduction of Marine Animals,* Vol. 1, Giese, A. C. and Pearse, J. S., Eds., Academic Press, New York, 1974, chap. 6.
Storer, T. I. and Usinger, R. L., *General Zoology,* 4th ed., McGraw-Hill, New York, 1965; 5th ed., with R. C. Stebbins and J. W. N. Nybakken, 1972.
Swan, E. F., Growth, autotomy, and regeneration, in *Physiology of Echinodermata,* Boolotian, R. A., Ed., Interscience Press, New York, 1966, chap. 17.
van Tienhoven, A., *Reproductive Physiology of Invertebrates,* W. B. Saunders, Philadelphia, 1968.
Vorontsova, M. A. and Liosner, L. D., *Asexual Propagation and Regeneration,* Pergamon Press, London, 1960.
Wardle, R. A. and McLeod, J. A., *The Zoology of Tapeworms,* University of Minnesota Press, Minneapolis, 1952.
Waterman, T. H., Ed., *The Physiology of Crustacea,* Vol. 1, Vol. 2, Academic Press, New York, 1960.
Williams, A. and Rowell, A. J., Brachiopod anatomy, in *Treatise on Invertebrate Paleontology,* (Part U), Vol. 1, Moore, R. C., Ed., Geological Society of America and University of Kansas Press, Lawrence, 1965.
Wilson, E. B., *The Cell in Development and Heredity,* Macmillan, New York, 1928.

INDEX

A

Acanthobellida, 127
Acanthocephala, 44, 57, 73, 124
Acanthocystis, 58, 119
Acarina, 87
Aceola, 123
Acrididae, 89, 128
Actinea, 67
Actinia, 122
Actiniaria, 122
Actinophrys, 42, 48, 119
Actinopoda, 119
Actinosphaerium, 58, 119
Actinula, 105
Activation, 8, 25—29, 97, 105
Activation fertilization, 25—28
Adult, 2, 8, 36, 38, 43, 105
Adult division, 44—45
Aeginopsis, 65
Aeolosomatidae, 84, 127
Agametes, 8, 9, 15, 30—31, 97, 105
Agametic, 19
Agamogenesis, 3, 30, 97, 105
Agamogony, 31, 44—45, 47, 97, 105
Agamont, 105
Agamy, 19, 29, 97, 105
Agnatha, 95, 130
Alcyonacea, 67, 122
Alcyonaria, 122
Allogamy, 25, 27, 97, 105
Allolobophora caliginosa, 84, 127
Allonais paraguayensis, 84, 127
Ameiosis, 15, 20, 105
Ameiotic, 20, 105
Ameiotic parthenogenesis, 12, 21—22, 29, 30, 37, 101, 105
Ameiotic reproduction, 14, 97
Amictic, 105
Amixis, 29, 97, 105
Amoebae, 105
Amoebida, 119
Amoebocyte, 105
Amoeboid, 105
Amoebulae, 105
Amphibia, 95—96, 130
Amphiblastula, 105
Amphigenesis, 25, 97, 105
Amphigonous, 105
Amphigony, 11, 25, 97, 105
Amphimictic, 19—20, 105
Amphimictic karyogamy, 25
Amphimictic reproduction, 23
Amphimixis, 3, 15, 19, 97, 105
Amphineura, 44, 79
Amphipholis squamata, 91, 129
Amphitoky, 29, 105
Anabiosis, 105
Ancestrula, 105
Androgenesis, 29, 97, 105
Anisogametes, 19, 25, 26, 97, 105
Anisogamous, 105
Anisogamy, 3, 25, 97, 105
Annelida, 45, 81—84, 126, 127
Annotation, 119—130
Anodonta, 80, 125
Anoplura, 128
Anthogenesis, 29, 97, 105
Anthomedusae, 121
Anthozoa, 44, 47, 67—68, 122
Aphididae, 89, 128
Aphis, 43
Aplacophora, 125
Apogamete, 30, 106
Apomictic, 19—20, 106
Apomixis, 3, 19, 25, 29, 97, 106
Apozygote, 30, 106
Appendicularia, 93
Arachnida, 45, 87, 127
Arcella, 58, 110
Archiannelida, 45, 84, 127
Architomy, 3, 14, 15, 32, 33, 34, 97, 106
Argeniaspis fuscicollis, 43
Argyrotheca, 78, 125
Armadillos, 47, 130
Arrhenotoky, 29, 97, 106
Arthropoda, 86—89
Articulata, 78, 125
Ascidiacea, 45, 47, 93—94, 129, 130
Asexual, 18, 20, 106
Asexual apomictic reproduction, 23
Asexual apomixis, 19
Asexual reproduction, 14, 26, 30—37, 46, 47, 97, 98, 106
    vs. sexual, 17—23
Asplanchnia, 43
Assembled bodies, 15, 98, 106
Asteroidea, 45, 90—91, 129
Astomata, 120
Athropoda, 127, 128
Atokal, 106
Atoke, 36, 106
Atubaria, 92, 129
Atubaria heterolopha, 92, 129
Aulophorus, 84, 127
Aurelis, 42
Autocopulation, 106
Autogamy, 2, 19, 38, 98, 106
Autolytus, 83, 126
Autolytus prolifer, 83, 126
Automictic, 106
Automictic parthenogenesis, 29, 106, 113
Automixis, 14, 18, 19, 22, 25, 27, 37, 38, 98, 106
Autotomized appendages, 9
Autotomy, 3, 9, 33, 34, 98, 106
Autozooid, 106
Aves, 96, 130

Avicularia, 106
Axial cells, 31
Axioms, 4—5
Axoblasts, 31, 106

B

Balanoglossus capensis, 93, 129
Basic Bisexual Reproduction, 13, 18, 19, 22, 23, 25, 47, 48, 98, 106
Bdelloidea, 73, 124
Beroe, 68, 122
Binary fission, 14, 32, 33, 98, 106
Biparental karyogamy, 23
Bisexual, 18, 20, 106
Bisexual reproduction, see also Basic Bisexual Reproduction, 1, 2, 14, 21—22, 98, 106
Bivalvia, 44, 50, 79—80, 125
Blastomeres, 8, 15
Blastostyle, 106
Boloceroides, 67, 122
Bonellia, 81, 126
Bougainvillea, 11
Brachiopoda, 44, 78, 125
Branchiopoda, 87, 127
Branchiura, 128
Bryozoa, 44, 76—77, 124, 125
Bud, 7, 8, 15, 36, 38, 43, 106
Budding, 9, 14, 23, 30, 33, 34, 35, 37, 42, 43, 47, 98, 99, 106
    multiple, see Multiple budding
Bugula, 43, 125

C

Calcarea, 61—62, 120
Calycophora, 122
Calyssozoa, 44, 76, 124
Calyx, 107
Cassiopeia, 67, 122
Catenulida, 70, 123
Cell constancy, 107
Cells, 38
Cephalochordata, 45, 94—95, 130
Cephalodiscus, 92, 129
Cephalodiscus sibogae, 92, 129
Cephalopoda, 44, 80, 125
Cercaria, 107
Cestoda, 42, 44, 68, 71—72, 123
Cestodaria, 44, 72, 123
Chaetoderma, 125
Chaetodermatidae, 79, 125
Chaetognatha, 45, 89, 128
Chaetonotoidea, 74, 124
Chaetopteridae, 126
Chaetopterus, 126
Chain of gonophores, 107
Chains, 35—36
Change of genome, 14

Cheilostomata, 125
Chelonia, 130
Child, 7
Chilopoda, 45, 88, 128
Chironomus grimmii, 89, 128
Chondrichthyes, 95, 130
Chonotricha, 60, 120
Chromosome aberrations, 20
Chrysaora, 67, 122
Chrysomonadina, 59, 119
Ciliata, 59, 60, 120
Ciliospore, 31, 107
Cirratulidae, 126
Cirripedia, 87, 88, 128
Cladocera, 42, 88, 127
Classification, 119—130
Clonal, 107
Clonal fertilization, 21, 23, 25—27, 44—45, 47, 99, 107
Clone, 2, 25, 107
Cnemidophorus, 96, 130
Cnidosporidia, 61, 120
Coccidae, 128
Coelenterata, 44, 64—68, 121, 122
Coeloplana, 68, 122
Coenurus bladder, 107
Coleoptera, 128
Colony, 7, 8, 107
Colony budding, 35, 42
Composite animals, 13
Composite egg, 15
Composite zygotes, 38
Configuration, 99
Conjugation, 9, 14, 18, 19, 22, 26, 37, 38, 107
Copepoda, 128
Copulation, 25, 107
Cormidium, 107
Craspedacusta, 66, 121
Crepidula fornicata, 79, 125
Crinoidea, 45, 90, 128
Crisia, 77, 125
Cross-fertilization, 2, 19, 21, 25—27, 99 107
Crossing over, 20
Crustacea, 45, 87—88, 127, 128
Cryptobranchus, 130
Ctenophora, 44, 68, 122
Ctenoplana, 68, 122
Cucumaria, 92, 129
Cunina, 65, 121
Cycles, 39—43, 107
Cyclostomata, 77, 125
Cydippida, 68, 122
Cynipidae, 89, 128
Cyst, 107
Cysticercus, 107
Cytogamy, 19, 38, 99, 107

D

Dactylopodalia, 75, 124
Dasypus novemcinctus, 43

Death, 8, 107
Demospongea, 120, 121
Demospongia, 61, 62
Dendrochirota, 92, 129
Dendrosomides, 60, 120
Deuterotoky, 29, 99, 107
Developmental cycle, 107
Developmental processes, 8
Diazona, 94, 130
Diazonidae, 94, 130
Dichogamy, 10, 99, 107
Dichotomous autotomy, 34, 99, 107
Dicyemennea schulszianum, 63, 121
Dicyemida, 63, 121
Didymozoonidae, 71, 123
Digenea, 71, 123
Dimorphism, 9
Dinoflagellata, 58—59, 119, 120
Dinoflagellates, 42
Dinophiloidea, 45, 85, 127
Dioecious, 10, 107
Dioecious cross-fertilization, 44—45
Dioecism, 10, 21, 99, 107
Dioecocestus, 71, 123
Diopisthoporus, 69, 123
Diploid, 107
Diploid parthenogenesis, 101, 113
Diploidy, 15, 25, 28, 29, 107
Diplophallus, 72
Diplopoda, 45, 88, 128
Diplozoon, 71, 123
Diporpa, 108
Diptera, 128
Dissogeny, 25, 99, 108
Dissogony, 25, 99, 108
Distaplia, 42, 94, 130
Distribution of Processes, 97—103
Diversity
  levels
    reproductive processes in a species, 44—49
    sequences or cycles, 39—43
    within a class, 49—55
Diversity factors, 5
Diversity in reproduction, 17
Diversity of individuality, 9
Diversity of reproduction, 1, 12
Division, 3, 15, 32—33, 99, 108
Dodecaceria caulleryi, 34, 126
Doliolida, 130
Doliolum, 94, 130
Domestic fowl, 130
Double budding, 35
Dual budding, 35
Dual embryo, 12
Dugesia, 69, 70, 123

E

Echiniscus, 85, 127
Echinococcus, 72, 123
Echinodermata, 45, 90—92, 128, 129
Echnoidea, 45, 91, 129
Echiurida, 81, 126
Echiuroidea, 45, 80—81, 126
Ectoprocta, 76—77
Egg, 108
Eleutheria, 66, 121
Elphidium, 58, 119
Embioptera, 89, 128
Embryo, 1, 2, 8, 42, 43, 108
Embryo colony, 12
Embryonic budding, 34—35
Embryonic division, 44—45
Embryonic fragmentation, 34—35
Embryonic multiplication, 34
Encrusting colony, 12
Encystment, 108
Endamoeba histolytica, 58, 119
Endogamy, 3, 25, 99, 108
Endogenous, 108
Endogenous budding, 35
Endomixis, 38, 99, 108
Endoprocta, 76, 124
Enteropneusta, 45, 93, 129
Entoprocta, 76
Epenthesis macgradii, 66, 121
Ephyrae, 36, 108
Epigamy, 37, 99, 108
Episode, 39, 40, 42, 108
Epitoke, 12, 36, 108
Epitoky, 3, 12, 33, 34, 36, 99, 108
Etheogenesis, 29, 99, 108
Eudorina, 59, 119
Eudoxy, 3, 33, 108
Eunicidae, 126
Eutardigrada, 85, 127
Eutely, 108
Exogamy, 3, 25, 27, 99, 108
Exogenous, 108
Exogenous budding, 35
Exogone gemmipara, 81, 126

F

Facultative parthenogenesis, 101
Fanworms, 126
Fasciola, 123
Fasciola hepatica, 71, 123
Female, 9, 11, 108
Fertilization, 20, 21, 25—28, 43, 108
  clonal, see Clonal fertilization, 27
  fraternal, see Fraternal fertilization, 27
  self-, see Self-fertilization, 27
Fission, 8, 9, 15, 32, 33, 37, 99, 109
  embryonic, see Embryonic fission, 2
  multiple, see Multiple fission, 2
Fissiparity, 3, 34, 99, 109
Fixed sequences, 48
Flagellata, 58, 59, 119, 120
Foetus, 7, 109
Foraminifera, 58, 119
Fragment, 8, 15

Fragmentation, 1, 8, 9, 14, 23, 30, 33, 37, 42, 43, 47, 99, 109
Fraternal fertilization, 21, 25—27, 44—45, 99, 109
Frustulation, 3, 33, 99, 109
Frustule, 109
Fundulus, 95, 130
Fungia, 68, 122
Fusion, 8, 19, 26, 38, 99, 109
Fusion of gametes, 20

G

Gall waps, 42
Gametes, 4, 7, 15, 18, 19, 25, 26, 33, 109
Gametic, 109
Gametic reproduction, 19, 100
Gametogenesis, 14, 21, 25, 100, 109
Gamogenesis, 11, 100, 109
Gamogony, 3, 9, 25, 100, 109
Gamont, 109
Gastrodes, 68, 122
Gastropoda, 44, 79, 125
Gastrotricha, 44, 74, 124
Gastrula, 42, 109
Gemmation, 34, 100, 109
Gemmiparity, 35, 100, 109
Gemmiparous stolonization, 36, 109
Gemmulation, 3, 7, 14, 30, 32, 100, 109
Gemmules, 8, 9, 15, 16, 32, 109
Geneagenesis, 29, 109
Genetic diversity, 4, 20, 28
Genetic ratios, 2, 23
Genome, 20—21, 109
Genome change, 14, 109
Genotype, 20, 109
Genotype change, 37
Germ ball, 42, 109
Germinal, 109
Glossary, 105—117
Gnathobdellida, 127
Gnathostomuloidea, 69, 123
Gonactinia prolifera, 67, 122
Gonad transplant, 10, 11, 100, 109
Gonads, 20, 26
Gonangium, 11, 46, 110
Gonochorism, 10, 21—22, 100, 110
Gonochorist, 18, 19, 110
Gonochoristic, 110
Gonochoronic individual, 10
Gonomery, 3, 21, 25, 27—29, 100, 110
Gonophore, 11, 110
Gonozooid, 110
Gordioidea, 44, 75—76, 124
Gorgonacea, 67, 122
Graphognathus leucoloma, 89, 128
Gregarinida, 60, 61, 120
Gymnolaemata, 42, 74, 77, 125
Gymnospores, 31, 110
Gynandromorph, 10, 11, 110
Gynandromorphism, 10, 110

Gynogenesis, 28, 29, 100, 110
Gyrocoelia, 72
Gyrodactylus, 70, 71, 123
Gyrodactylus elegans, 70, 123

H

Hadromerina, 120
Haemosporidia, 60—61, 120
Haliclystus, 122
Haploid, 110
Haploid gametes, 15
Haploid parthenogenesis, 101, 113
Haploidy, 25, 110
Haplosclerina, 121
Haplozoon, 59, 119
Hectocotyle, 7
Hectocotylus, 110
Heliactis, 67, 122
Heliozoa, 119
Hemimixis, 38, 100, 110
Hemizygoid, 110
Hemizygoid parthenogenesis, 29, 101, 110
Hermaphrodite, 10—12, 18, 19, 47, 110
Hermaphroditic, 110
Hermaphroditic cross-fertilization, 44—45
Hermaphroditic individual, 10
Hermaphroditism, 10, 22, 44—45, 100, 110
Herpobdella, 84, 127
Heterodera, 75, 124
Heterogamete, 110
Heterogamy, 25, 100, 110
Heterogenesis, 110
Heterogenotypic, 110
Heterogony, 110
Heterogyny, 110
Heterospermic, 110
Heterospermy, 110
Heterostephanus, 66, 121
Heterotardigrada, 85, 127
Hexactinellida, 62, 120
Hibernacula, 3, 16, 32, 100
Hibernaculum, 110
Hirudinea, 45, 84, 127
Hologamete, 19, 25, 26, 111
Hologamous, 111
Hologamy, 3, 8, 9, 14, 18, 19, 22, 25, 26, 37, 38, 100, 111
Holothurioidea, 45, 92, 129
Holotricha, 60
Homoptera, 128
Homo sapiens, 47, 48
Homospermic, 111
Homospermy, 111
Homozygous, 20
Humans, 47, 130
Hyatid cyst, 111
Hydra, 48, 65, 66, 121
Hydranth, 11, 46, 111
Hydrocaulus, 111
Hydrozoa, 42, 44, 55, 64—66, 121, 122

Hymenoptera, 42, 89, 128
Hypodermic injection, 11
Hypodermic insemination, 111
Hypogenesis, 100, 111
Hypolytus, 66, 121

I

Ibla, 87, 128
Icerya, 89
Icerya purchasi, 128
Identical twinning, 34
Inarticulata, 78, 125
Inbreeding, 25, 27, 100, 111
Individual, 8, 10, 12, 17, 20, 111
Individualistic colony, 12
Individuality, 7—9, 13, 111
Infusorigen, 31, 111
Insecta, 42, 45, 47, 53, 88—89, 128
Insemination, 14, 23, 25, 27, 28, 111
Insemination by spermatophore, 111
Intersex, 10, 11, 111
Intersexuality, 10
Isogametes, 18, 19, 25, 26, 111
Isogamous, 111
Isogamy, 3, 18, 19, 25, 100, 111
Isometra vivipara, 90, 129
Isopoda, 87, 128

K

Kamptozoa, 76
Karyogamy, 15, 18—20, 21, 25, 26, 28, 100, 111
Kinorhyncha, 44, 74—75, 124

L

Lacerta saccicola, 96, 130
Larva, 1, 2, 8, 38, 42, 43, 111
Larvacea, 45, 93, 129
Larval budding, 35
Larval division, 44—45
Lernaeophrya capitata, 60, 120
Life cycle, 39, 107, 111
Limnodrilus udekemianus, 84, 127
Linckia, 91, 129
Lineus, 42, 43, 123
Lineus sanguineus, 72
Lingula, 78, 125
Liriope, 65
Lobata, 68
Lobosa, 42, 58, 119
Loxosoma, 76, 124
Loxosomatidae, 76, 124
Lucernariidae, 67, 122
Lumbriculidae, 127
Lumbriculus variegatus, 84, 127
Lumbricus trapezoides, 84, 127
Lymnaea, 79, 125

M

Macrodasyoidea, 74, 124
Madreporaria, 67, 122
Maenidae, 95, 130
Malacostraca, 128
Male, 9—11, 111
Mammalia, 96, 130
Margellium, 65, 121
Mating types, 11
Medusae, 11, 46, 111
Medusoids, 111
Meiosis, 4, 18, 20, 21, 25, 30, 100, 111
Meiotic, 20, 111
Meiotic parthenogenesis, 14, 21—22, 28, 30, 37, 101, 111
Membranipora, 77, 125
Mendelian ratios, 47
Mendel's laws, 4
Mermis, 75, 124
Merogametes, 111
Merogony, 3, 29—31, 100, 111, 112
Merostomata, 45, 50, 86—87, 127
Merozoites, 31, 33, 112
Mesorhabditis belari, 75, 124
Mesozoa, 44, 63—64, 121
Metacercaria, 112
Metagenesis, 46, 112
Metamorphosis, 36, 112
Metridium, 67, 122
Mexican axolotl, 130
Miastor, 89, 128
Microhydra, 66, 121
Micromalthus, 89, 128
Mictic, 112
Miracidium, 112
Mitotic, 112
Mixing of genomes, 23
Mixis, 19, 25, 101, 112
Moerisia, 66, 121
Mollusca, 44, 78—80, 125
Monaxonida, 121
Monoblastozoa, 44, 64, 121
Monodisk, 36, 112
Monodisk strobilation, 36
Monoecious, 112
Monoecious species, 10
Monoecism, 10, 22, 101, 112
Monogenea, 70, 71, 123
Monogononta, 73—74, 124
Monogony, 30, 101, 112
Monoplacophora, 78—79, 125
Monotomy, 33, 101, 112
Multicellular reproductive bodies, 30
Multiceps, 72, 123
Multiple budding, 2, 35
Multiple fission, 2, 14, 31—33, 101, 112
Multiple gonophores, 36
Multiple pathways, 46
Multiple segment sequences, 48—49
Multiplication, 13, 14, 34, 112
Multiplication of individuals, 18
Multiplicative autotomy, 33

Mutation, 20, 112
Mychogamy, 3, 27, 101, 112
Myxine glutinosa, 95, 130
Myxosporidia, 42, 61, 120
Myzostomida, 45, 51, 81, 126
Myzostomum, 81, 126

N

Naididae, 127
Nais, 84, 127
Narcomedusae, 66, 122
Nectonematioidea, 76, 124
Nematoda, 44, 75, 124
Nematogen, 31, 112
Nematomorpha, 75—76, 124
Nemertodermatida, 69, 122
Neodasys, 74, 124
Neoteny, 112
Nephtyidea, 126
Nephtys, 83, 126
Nereidae, 126
Nereis irrorata, 83
Neuters, 9—12
Noctiluca, 59, 120
Nonbisexual reproduction, 2
Nongametic, 19—20, 112
Nongametic apomixis, 29
Nongametic reproduction, 14
Nuclear reorganization, 9, 14, 18, 19, 26, 37, 38, 101, 112
Nuclearia, 58, 119
Nuclei, 19, 26
Nuda, 68, 122
Nurse cells, 112

O

Obelia, 11 40, 43, 46, 65
Obligate parthenogenesis, 49
Obligate sequences, 41
Odonata, 89
Offspring, 20
Oikopleura dioica, 93, 129
Oligochaeta, 42, 45, 84, 126, 127
Onchomiracidium, 112
Oncosphere, 112
Onychophora, 45, 86, 127
Oocyst, 113
Oogamy, 25, 101, 113
Oogenesis, 25, 101, 113
Ookinete
Opalinida, 59, 60, 120
Ophiacantha vivipara, 91
Ophiactis, 91, 129
Ophiuroidea, 45, 91, 129
Opisthobranchia, 79, 125
Orthonectida, 63—64, 121
Orthoptera, 89, 128

Osteichthyes, 95, 130
Ostracoda, 87, 128
Ostrea, 80, 125
Outbreeding, 4, 19, 20, 22, 27, 101, 113
Ova, 9, 15, 25
Oviparous, 113
Ovotestis, 101, 113
Ovoviviparous, 113
Ovum, 8, 26, 30, 113

P

Paedogamous, 113
Paedogamy, 28, 38, 101, 113
Paedogenesis, 113
Palintomy, 31, 101, 113
Pallial, 113
Pallial budding, 35
Palolo siciliensis, 83, 126
Pandorina, 59, 119
Pansporoblast, 113
Paramecium, 42, 47, 60, 120
Paramoeba, 58, 119
Parareproductive, 44—45, 113
Parareproductive processes, 13, 37—38
Paratomy, 3, 15, 32—35, 101, 113
Parent genotypes, 20
Parorchis acanthis, 70—71, 123
Parthenogamy, 113
Parthenogenesis, 2, 3, 9, 11, 12, 14, 15, 17, 20—23, 25, 26, 28—30, 37, 43—47, 101, 113
   obligate, see Obligate parthenogenesis, 49
Parthenogenetic, 113
Patellina, 58, 119
Pathways, 46, 50—55
Patrogenesis, 29, 101, 113
Pauropoda, 45, 88, 128
Pecten, 79, 125
Pedal laceration, 3, 33, 101, 113
Pediculus, 128
Pelagia, 67, 122
Peltogasterella, 87, 128
Pentastomida, 45, 86, 127
Peripatus, 47
Peritricha, 60, 120
Phagocata verlata, 70, 123
Phalangium opilio, 87, 127
Pharyngobdellida, 127
Phialidium gastroblasta, 121
Phoronida, 44, 77—78, 125
Phoronis muelleri, 77, 125
Phoronis ovalis, 78, 125
Phylactolaemata, 47, 76—77, 124, 125
Phyllochaetopterus, 81, 83, 126
Physalia, 8
Phytomonadina, 59, 119
Placozoa, 64, 121
Planaria, 70, 123
Planctosphaeroidea, 93, 129

Planula, 113
Plasmodium, 8, 9, 31, 113
Plasmodium, 60—61, 120
Plasmodium vivax, 43
Plasmogamy, 29, 102, 113
Plasmogony, 2, 3, 25, 28, 29, 102, 113
Plasmotomy, 3, 33, 102, 113
Plastogamy, 38, 102, 114
Platasterias, 90, 129
Platidia, 78, 125
Platyctenea, 68, 122
Platyhelminthes, 44, 68—72, 122, 123
Platynereis dumerilii, 82, 83
Platynereis massiliensis, 81—82
Platynereis spp., 126
Plecoptera, 128
Plerocercoid, 114
Podocysts, 3, 16, 32, 102, 114
Pogonophora, 45, 89—90, 128
Polar body, 28, 114
Polar body fertilization, 21, 25, 28, 29, 102, 114
Polychaeta, 42, 45, 47, 81—83, 126
Polyclinidae, 42, 94, 130
Polydisk, 114
Polydisk strobilation, 36
Polyembryony, 1—3, 8, 9, 23, 33—35, 42, 44—45, 49, 102, 114
Polygenotypic, 114
Polymorphic, 114
Polyp, 11, 12, 36, 114
Polyplacophora, 125
Polyploidy, 114
Polypoids, 114
Polyspermy, 114
Polystoma integerrium, 70, 123
Polyzoa, 76—77
Porifera, 61—63, 120, 121
Prebuds, 35, 102, 114
Preproductive, 114
Prereproductive processes, 14, 38
Priapuloidea, 44, 75, 124
Process, 3, 39
Progenesis, 29, 102, 114
Proglottids, 7, 9, 36, 114
Prosobranchia, 79, 125
Protamoeba primitiva, 33, 57, 119
Protandric, 114
Protandrous, 114
Protandry, 10, 102, 114
Proteomixida, 119
Protociliata, 59, 120
Protodrilus, 84, 127
Protogamy, 114
Protogenesis, 35, 102, 114
Protogynous, 114
Protogyny, 10, 102, 114
Protohydra, 121
Protozoa, I: 42, 44, 47, 57—61, 119, 120
Pseudo-eggs, 31, 114
Pseudogamy, 2, 3, 14, 22, 25, 28, 29, 37, 102, 114
Pseudophyllidea, 72, 123
Pseudo-planulae, 114

Pseudovum, 114
Pterobranchia, 92, 129
Pulmonata, 79, 125
Pumilus, 78, 125
Pycnogonida, 87, 127
Pyloric budding, 15, 35, 114
Pyrosoma, 94, 130
Pyrosomata, 94, 130

R

Radiolaria, 58, 119
Raja batis, 130
Recombination, 4
Rediae, 114
Reduction bodies, 9, 15, 32, 102, 114
Reduction divisions, 30
Regeneration, 15, 115
Reproduction, 11—14, 17
  animals by class
    acanthocephala, 73
    annelida, 81—84
    brachiopada, 78
    bryozoa, 76—77
    calyssozoa, 76
    cephalochordata, 94—95
    chaetognatha, 89
    coelenterata, 64—68
    ctenophora, 68
    dinophiloidgea, 85
    echinodermata, 90—92
    echiuroidea, 80—81
    enteropneusta, 93
    gastrotricha, 74
    gordioidea, 75—76
    kinorhyncha, 74—75
    mesozoa, 63—64
    mollusca, 78—80
    monoblastozoa, 64
    myzostomida, 81
    nematoda, 75
    nematomorpha, 75—76
    onychophora, 86
    pentastomida, 86
    phoronida, 77—78
    placozoa, 64
    planctosphaeroidea, 93
    platyhelminthes, 68—72
    pogonophora, 89—90
    porifera, 61—63
    priapuloidea, 75
    protozoa, 57—61
    pterobranchia, 92
    rhynchocoela, 72
    rotifera, 73—74
    sipunculoidea, 80
    tardigrada, 85
    tunicata, 93—94
    vertebrata, 95—96
  bisexual, see Bisexual reproduction, 1

nonbisexual, see nonbisexual reproduction, 2
Reproductive behavior, 14
Reproductive bodies, 14—16, 25, 115
Reproductive cycle, 115
Reproductive episodes, 40
Reproductive pathways, 50—55
Reproductive processes, see also Specific process
   activation, 25—28
   asexual reproduction, 30—37
   fertilization, 25—28
   in a species, 46—49
   parthenogenesis, 28—30
Reproductive sequences, see also Sequences, 41, 46
Reptilia, 96, 130
Resting buds, 16, 35, 102, 115
Rhabditis, 75
Rhabdocoela, 123
Rhabdopleura, 92
Rhizocephala, 87, 88, 128
Rhizopoda, 57, 119
Rhizostomeae, 122
Rhombogen, 115
Rhynchobdellida, 127
Rhynchocoela, 44, 47, 72, 123
Ricinulei, 87
Rotifera, 124

S

Sabella, 126
Sabella spallanzanii, 83
Sabellidae, 81, 126
Saccosomatida, 81, 126
Sagartia, 67, 122
Sagitta, 89, 128
Salinella salve, 64, 121
Salpa, 94, 130
Salpida, 94, 130
Sappinia, 119
Sappinia diploidea, 58, 119
Sarcodina, I: 47, 57—58, 119
Scalpellum, 87, 128
Scaphopoda, 44, 80, 125
Schistosomatidae, 71, 123
Schizocladium, 66, 121
Schizogamy, 37, 102, 115
Schizogenesis, 34, 102, 115
Schizogony, 3, 33, 102, 115
Schizometamery, 33, 34, 102, 115
Schizont, 33, 115
Schyphistoma, 115
Schyphozoa, 44, 47, 122
Scissiparity, 3, 34, 102, 115
Scleractinia, 122
Sclerospongea, 62—63, 121
Scolex, 36
Scyphistoma, 32, 34, 36, 46
Scyphozoa, 66—67
Secondary budding, 35, 115
Segment, 39, 44—45

Seisonidea, 73, 124
Self-fertilization, 2, 3, 21—23, 25, 27, 44—45, 102, 115
Self-fertilizing hermaphrodites, 47
Semaeostomeae, 67, 122
Sequences, 39—43, 46—49, 115
Sequential hermaphroditism, 11
Serranidae, 95, 130
Sex, 9, 17, 23, 115
Sex arrangement, 11
Sex reversal, 11, 115
Sexual, 18, 20, 115
Sexuality, 4, 9—12, 17, 115
Sexual reproduction, 2, 12, 17, 21, 37, 102, 115
   vs. asexual, 17—23
Single segment sequences, 46—48
Siphonophora, 65, 122
Sipunculoidea, 45, 80, 126
Solenium, 115
Solenogastres, 44, 79, 125
Somasteroidea, 90, 129
Somatic, 115
Somatic fertilization, 25, 28, 38, 102, 115
Somatic fission, 33, 102, 115
Sorites, 3, 16, 32, 102, 115
Sparganum prolifer, 123
Sparidae, 95, 130
Species, 115
Sperm, 115
Spermaries, 115
Spermatogenesis, 25, 102, 116
Spermatophore, 7, 9, 116
Spermatozoa, 25, 116
Sphaeriidae, 125
Sphyranura, 70, 123
Sponges, 9, 61—62
Spongillidae, 61, 62, 121
Spore-cell, 116
Spore formation, 31, 116
Spores, 30—31, 33, 102, 116
Sporocyst, 116
Sporogamy, 31 102, 116
Sporogony, 3, 14, 21, 22, 31, 37, 102, 116
Sporozoa, 60—61, 120
Sporozoites, 31, 33, 102, 116
Sporulation, 31, 102, 116
Statoblasts, 3, 9, 16, 32, 102, 116
Statocyst, 32, 116
Staurogamy, 3, 27, 102, 116
Stauromedusae, 122
Stenostomum, 70, 123
Stereogastrula, 116
Stolon, 36, 37, 43, 116
Stolon budding, 35
Stolonifera, 67, 122
Stolonization, 3, 35, 47, 102, 116
Strepsiptera, 89, 128
Strobila, 36, 116
Strobilation, 2, 3, 9, 14, 33—37, 42, 43, 103, 116
Stromatoporoidea, 62
Stylonychia, 60, 120
Subitaneous, 116

Successional polyembryony, 3, 34, 103, 116
Suctoria, 60, 120
Swarmers, 31, 116
Syllidae, 83, 126
Syllis, 81, 83, 126
Syllis gracilis, 83, 126
Syllis hyalina, 83, 126
Syllis ramosus, 83, 126
Syllis vittata, 83
Symbranchidae, 95, 130
Symphyla, 45, 88, 128
Synaptida, 92, 129
Synchronogamy, 10, 103, 116
Syncytium, 116
Syngamy, 14, 19, 21, 25—27, 103, 116
Syngamy-plus-karyogamy, 25
Syngony, 10, 103, 116
Synkaryon, 21, 26, 116
Syntomy, 37, 103, 116, 117
Syzygy, 38, 103, 117

T

Taenia crassiceps, 72, 123
Taenioidea, 72, 123
Tardigrada, 45, 85, 127
Telosporidia, 120
Temnocephaloidea, 70, 123
Tentaculata, 68, 122
Testacida, 119
Tethya maza, 62, 121
Thaliacea, 45, 94, 130
Thelytoky, 29, 103, 117
Theory of Recapitulation, 4
Tomiparity, 3, 33, 34, 103, 117
Trachydermon raymondi, 79, 125
Trachylina, 121
Trematoda, 44, 54, 70—71, 123
Trichinella spiralis, 75, 124
Trichoplax, 64
Trichosphaerium, 58, 119
Tricladida, 69, 123
Tridideminum, 94, 130
Trochophore, 117
Trophozoites, 117
Trypanosillis, 83, 126
Trypanosoma, 120
Trypanosoma lewisii, 59, 120
Trypetesa, 88, 128
Tubificidae, 127
Tunicata, 45, 93—94, 129, 130
Turbellaria, 44, 69—70, 123
Twinning, 103, 117
Tylorrhynchus, 126

Tylorrhynchus heterochaetus, 83
Typhloplanidae, 69

U

Unicellular reproductive bodies, 25
Unisexual, 18, 103, 117
Unisexuality, 117
Unity of all life, 3
Urocystidium, 72, 123
Urocystis, 72, 123

V

Vampyrella, 58, 119
Variable sequences, 48
Vermiform bodies, 117
Vertebrata, 45, 51, 52, 95—96, 130
Vertebrates, 1
Viviparity, 117
Viviparous, 117
Volvox, 8, 59, 119
Vorticella, 60, 120

W

Winter buds, 117
Winter eggs, 117

X

Xenotrichula, 74, 124
Xenoturbellida, 69, 122

Z

Zeppelina, 42, 83
Zeppelina monostyla, 82, 83, 126
Zoantharia, 122
Zoogamy, 11, 103, 117
Zooids, 36, 117
Zufala, 72
Zygogenesis, 11, 103, 117
Zygogenetic, 117
Zygoid, 117
Zygoid parthenogenesis, 29, 101, 117
Zygoidy, 28, 117
Zygote, 2, 4, 7, 8, 21, 27—28, 38, 42, 117
Zygote nucleus, 21